I0112447

CALIFORNIA
SNAKES
AND HOW TO FIND THEM

Common Sharp-tailed Snake.
Photograph by Spencer Riffle

CALIFORNIA SNAKES

AND HOW TO FIND THEM

EMILY TAYLOR

H

HEYDAY

50

Berkeley, California

Copyright © 2024 by Emily Taylor

All rights reserved. No portion of this work may be reproduced or transmitted in any form or by any means, electronic or mechanical, including photocopying and recording, or by any information storage or retrieval system, without permission in writing from Heyday.

Library of Congress Cataloging-in-Publication Data
Names: Taylor, Emily (Lecturer in biological sciences), author.
Title: California snakes and how to find them / Emily Taylor.
Description: Berkeley, California: Heyday, [2024]
Identifiers: LCCN 2023030203 (print) | LCCN 2023030204 (ebook) | ISBN 9781597146340 (paperback) | ISBN 9781597146357 (epub)
Subjects: LCSH: Snakes—California. | Snakes—California—Identification. | Colubridae—California. | Colubridae—California—Identification.
Classification: LCC QL666.O6 T393 2024 (print) | LCC QL666.O6 (ebook) | DDC 597.9609794—dc23/eng/20231130
LC record available at https://lccn.loc.gov/2023030203
LC ebook record available at https://lccn.loc.gov/2023030204

Cover Art: Marisa Ishimatsu. Common Garter Snake
Cover and Interior Design: Debbie Berne

Published by Heyday
PO Box 9145, Berkeley, California 94709
(510) 549-3564
heydaybooks.com

Printed in East Peoria, Illinois, by Versa Press, Inc.

10 9 8 7 6 5 4 3 2

To Mom & Pop,
who brought me up to believe
I could become anything.
I don't think they meant a snake biologist.
But here we are.

CONTENTS

Panamint Rattlesnake.
Photograph by Marisa Ishimatsu.

◀ California Mountain Kingsnake.
Photograph by Marisa Ishimatsu.

Sidewinder.
Photograph by Zeev Nitzan Ginsburg.

Long-nosed Snake.
Photograph by Marisa Ishimatsu.

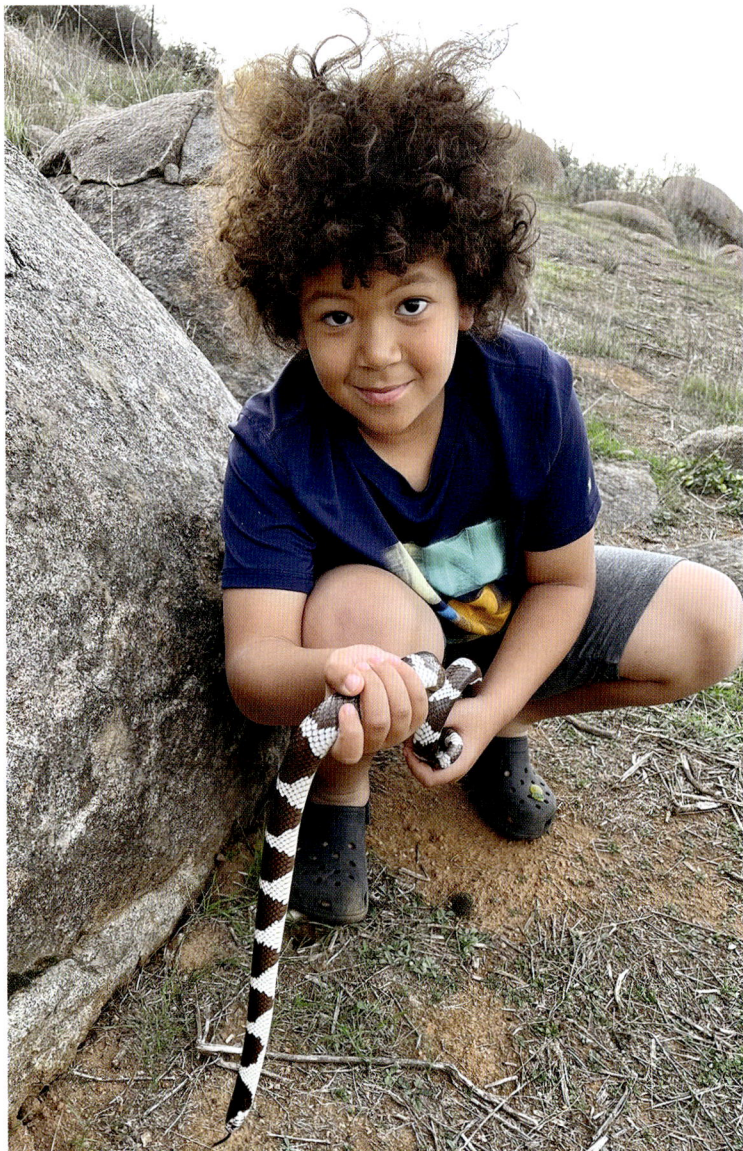

A child gently handles a California Kingsnake.

Cory Lindsay

PREFACE

This book is for snake lovers and snake lovers-to-be. Few animals capture our imaginations like snakes do, but their reputation is as forked as their tongues. For many people, snakes are scary, gross, and even considered bad omens. For others, like me, snakes represent grace, beauty, and resilience—they are just as fantastic as any beasts that Harry Potter encountered in the wizarding world. I may be just about as Slytherin as they come, but there will be no foolish wand-waving or silly incantations in this book: Learning about California snakes and how to find them requires practice, persistence, and luck. In this book, I coach you on what you need to develop snake-hunting skills and cheerlead you on being persistent enough to find the snakes you seek. But ultimately, we will have to leave luck up to the "herping gods . . . ," which my students and I jokingly credit on our successful field trips (and blame for less successful trips).

My love affair with snakes began twenty-five years ago on a class field trip in college when my herpetology professor handed me a California Kingsnake he had found under a log. My vision went dark around the edges as I stared, mesmerized, at the shiny black and white coils of muscle writhing in my hands. From then on, it was a rare log, rock, or piece of tin I passed without looking underneath it. I was hooked. I now spend much of my time spreading the good word about California snakes in every way I can by teaching my own herpetology students, unlocking the secrets of snakes in my research at Cal Poly, promoting ways to live safely in areas with rattlesnakes, and sharing snake stories everywhere I go.

While my goal for you, reader, is to help you find, identify, and observe California snakes out in the field, notably, this book is not a true field guide in the sense that I do not provide detailed information like scale counts, nor do I show range maps. In some cases, I merge closely related species into a single species account. You can find detailed technical information about each snake species in various field guides and websites, including the wonderfully comprehensive and convenient website www.CaliforniaHerps.com, and I have the full list of scientific references I drew from archived at my website as well at EmilyTaylorScience.com.

In addition to helping you learn how to find, observe, and identify the snakes around you, I want you, the budding snake enthusiast, to learn how to interact with wildlife in a responsible manner. Snakes have been unfairly portrayed in the media and persecuted mercilessly for ages. Even people who love snakes can sometimes harm them by destroying their habitat in an effort to catch them or by handling them roughly in pursuit of the perfect photograph. In this book, you will not just learn how to find snakes in the wild, but how to leave no trace behind in your search for these majestic beasts. Rattlesnakes should always be left alone, and standing back and watching even harmless snakes is the key to seeing once-in-a-lifetime behaviors like snakes catching prey or fighting with one another over a mate. That said, part of the joy of herping is approaching snakes with knowledge and caution in order to hold these striking creatures in your hands and observe them up close. As herpetologist Sam Sweet wrote, "Field herpetology cannot be (and should not be) reduced to birdwatching."

Snakes native to California have coexisted with California's Indigenous people for millennia. Rattlesnakes in particular have been significant animals in the daily lives of people in California, both now and in the deep past. Anthropologists have recorded diverse practices and beliefs among Indigenous groups with respect to rattlesnakes; for example, the use of rattlesnake fat to treat "consumption" (possibly tuberculosis) by the Northern Californian Pomo people, and the belief among the Shoshone, Luiseño, and Diegueño that rattlesnake bites are punishments for dereliction of religious duty. I acknowledge and pay homage to the fact that I write this book from the homeland of the Indigenous Te'po'ta'ahl, or "People of the Oaks," named by missionaries as the Antoniano division of the Salinan tribe, a group that is currently seeking official recognition from the Bureau of Indian Affairs.

California is famous for many reasons, but chief among these is its diversity. Diversity of cultures, of people, of wildlife, of climates, of crops, of wines . . . you name it. Californian snakes are no different. We have boas, we have vipers, we have blindsnakes, we have a legion of snakes from a large and diverse family called the Colubridae, and in recent years we even have had a few wayward sea snakes find their way to our shores. Fellow snake admirer, I am excited to be your guide on this journey into the wonderful world of Californian snakes.

Western Terrestrial Garter Snake. *Photograph by Marisa Ishimatsu.*

INTRODUCTION

What Are Snakes?

The most common answer to the question "What are snakes?" is "legless reptiles." But being a snake goes beyond just being legless (see page 24). So, what is a snake?

Animals are organisms with muscles and nerves, which are tissues lacking from other life like plants and fungi. Reptiles are animals with vertebral columns, eggs with certain membranes to protect the embryos outside of the water, and without fur or feathers. "Herps" is a broader category that refers collectively to amphibians and reptiles, which are often lumped together because they visually appear similar (think lizard and salamander) or because they share important biological features, like obtaining their body heat from the environment. It therefore follows that herpetology is the study of amphibians and reptiles, "herping" is going out looking for amphibians and reptiles, and a "herper" is one who goes herping (you and me!).

Reptiles include turtles, alligators and their relatives, a lizard-like reptile from New Zealand called the tuatara, and a large and diverse group consisting of lizards, snakes, and their relatives. Within lizards, legs have been lost on numerous occasions over evolutionary time. One lineage that lost its legs evolved into thousands of related species; this is the group that we call snakes today. Other lizards that independently lost their legs are not called snakes but are called legless lizards, and they have other characteristics that snakes don't have, like eyelids and external ears.

Today there are about 4,000 species of snakes that have been described by scientists, and many more that have yet to be discovered. Snakes live in most habitats on the planet, from the seas to the peaks of mountains, from tropical rainforests to deserts, from deep in the soils to high in the canopies. About fifty of these snake species live in California, and a few are endemic, meaning that they are native to our state and only occur here. A few others are introduced, released here by people either on purpose or by accident, and they became established. These non-native species can be a serious problem because they can spread disease, eat or displace native wildlife, or otherwise impact our biodiversity. This is one major reason for learning to identify Californian snakes: You can help scientists monitor the health of populations of both native and non-native species by posting your observations to websites like iNaturalist.

California Is a Perfect State for Snakes

California's biodiversity is legendary. Our beautiful state boasts over 6,500 varieties of plants, many of which occur nowhere else. Nearly half of the bird and mammal species and a third of the reptiles that occur in the United States can be found in California. Snakes are no exception—California has a wide array of snake species, and in some areas, snakes reach impressive biomasses due to their sheer abundance.

Why is California so special for snakes and other wildlife? Part of the answer is obvious: California is a huge state. Driving from the northern border to the southern border of California would be like driving through eleven East Coast states, from Maine to North Carolina. By virtue of its huge size, California houses the space to encompass many kinds of snakes. The influence of the sea is important in shaping complex microclimates that have been amenable to the evolution of a striking diversity of organisms. California boasts multiple habitat types, from the vast Mojave Desert to the majestic Sierra Nevada. Finally, California's biodiversity is extended by the presence of the Baja California peninsula to the south, which originated as a piece of land on the West Coast of mainland Mexico, broke off, and nudged slowly northward as a huge island for millions of years, then slammed into Southern California about 100 million years ago, forming the San Jacinto and Santa Rosa Mountains in the process. All the organisms that had been independently evolving on the island could now move northward, and California species could move southward. This, plus its warm and hospitable climate, is a main reason why Southern California in particular has such high reptile biodiversity even in the face of extreme urbanization.

California's vast parks and public lands are particularly welcoming playgrounds for the snake seeker. In many other states, hopeful herpers are blocked by locked gates at every turn. In California, you can purchase a fishing license that allows you to capture and admire snakes before releasing them into the wild, or even to keep them in captivity if you so desire. But watch out—most parks prohibit capturing wildlife on their grounds, even just to look at. I strongly recommend leaving wild snakes in the wild and sating your desire for a pet snake by buying a captive-bred snake at a reptile expo. The search for wild snakes is a joyful adventure that's its own reward. You'll go to beautiful places and explore trails less traveled, because that is where the snakes hang out.

People and Snakes—A Complicated Relationship

Most people have strong opinions about snakes. Whether they fear and hate them, or are fascinated by them, no one's going to shrug when you ask them what they think about snakes. Relationships with snakes are deeply ingrained in human history and cultures from around the world. In some religions, snakes are worshipped as deities, and in others they are vilified, and even when ideas of the sacred and demonic aren't involved, snakes evoke strong reactions. Today, ophidiophobia (fear of snakes) is the most common phobia in the world, yet the reptile house at your local zoo is among its biggest attractions.

Why are we so opinionated, one way or the other, about snakes? The reason undoubtedly lies deep inside our brains. As primates, our evolutionary ancestors co-evolved with snakes for millions of years. Primates were, and indeed still are, hunted by large, constricting snakes. Those individuals who could detect a venomous snake curled up in leaf litter or a large python hiding in a tree were less likely to be envenomated or eaten and more likely to pass on genes to future generations. In this way, the ability to detect snakes shaped the neural circuitry of primate brains and vision, with the ability to rapidly pick a snake out of a crowd still going strong in humans today.

But is *fear* of snakes innate? No, at least not really. The simple observation that most children are fascinated by snakes is evidence that we are not born fearful of snakes. Rather, fear is mostly learned. Children are little sponges that soak up the world around them, watching and emulating grownups carefully. When adults exhibit fear or disgust of snakes, kids are quick to copy it. When they see scary footage of a rattlesnake striking at their hero hosts on *Animal Planet*, the snake becomes the villain.

Because of the way our evolutionary history with snakes has shaped our brains, negative or positive associations with snakes are made much more rapidly and last much longer than for other animals. Kids who are exposed to snakes only from sensationalist

John Perrine

Two sisters show off a Gopher Snake and California Kingsnake they found before releasing them where they caught them.

nature shows are likely to develop fear or disgust toward snakes, and possibly toward other animals, too. Even worse, I shudder to think what becomes of a child's empathy for wildlife after watching adults torture and kill snakes at annual "rattlesnake roundups" that occur in some states. In contrast, taking kids to zoos and outreach events where they can meet snakes in a positive setting can have a lifelong effect on the way they see animals and nature. Even better, take a kid out herping and help them find their first wild Gopher Snake. Hanging out with snakes is a surefire way to raise an eager naturalist. Once they see their first, they want more.

Most California snakes are harmless to people. However, almost every inch of California is (or was, prior to urbanization) inhabited by one of the seven species of rattlesnakes that occur here. Given the huge diversity of rattlesnake stories, myths, and uses among Southern Californian Indigenous groups, hundreds of years ago these snakes must have played a huge role in people's day-to-day lives. While Benjamin Franklin helped establish rattlesnakes as a symbol of the grit of Americans in the Revolutionary War, the westward migration of Euro-American colonizers into what is now California was certainly accompanied by many snakebites that resulted in pain, disability, and death. This threat was a major factor shaping the relationships between people and snakes until the development of antivenom in the twentieth century. Even today when snakebites are less common and death from bites is exceedingly rare, snakebites are a major medical issue due to risk of permanent injury, not to mention the incredibly high cost of medical treatment. In this book, you will learn that rattlesnakes are gentle creatures that only bite to defend themselves against a perceived threat, which could include someone trying to kill them or accidentally stepping on them. Therefore, interacting with rattlesnakes is potentially high risk, and they should only be admired from a safe distance away.

If you are reading this book, you have chosen to learn more about snakes. Perhaps it's because you already love snakes, or maybe someone gifted you this book so that it might be your gateway into a life of fascination with the natural world. Read it, go herping, use the book to learn about snakes, and keep herping.

What's in a Snake's Name?

Taxonomy, the art of naming organisms and sorting out their relationships to each other, is often attributed to Carl Linnaeus, who developed a system of scientific nomenclature in the 1700s that is still in use today. However, people have been naming animals, snakes included, since long before that. For example, the Chumash people, who have inhabited the California Central Coast for many centuries, call the Western Rattlesnake that occurs here the "Xshap." Using that name in conversation conveys meaning not just about the snake's identity, but potentially about safety since that species is venomous. In other words, the names that we give snakes are part of our language and communication systems. Names literally allow us to know what we are talking about. No matter which language we speak, the names we use have to be understood by others in our communication system, so that we'll all hold the common idea of that animal collectively and understand what each person is referring to clearly.

In its early days, taxonomy was based mainly on appearance. Species were delineated by features like size, shape of the head, number and arrangement of scale rows, and other features. Nowadays, this has been augmented by molecular data. Phylogenetic systematics is the study of the evolutionary relationships among organisms, and often relies on morphology (physical characteristics) and molecules (DNA). Systematists studying the evolution of snakes sometimes find evidence that what we previously thought was a single species is actually two or more species, and they publish a paper *splitting* the species. Sometimes they find evidence that two or more species are just one, and they publish a paper *lumping* them into a single species. Taxonomy is inherently a subjective science, and systematists regularly argue it out in publications and at conferences. All of this explains why the taxonomy I use might be different from what you find in an older book and might also change in the future based on the latest science. In this book, I follow the convention of using names that have been proposed, have had time to be vetted by other scientists, and (so far) have withstood the test of time.

Snake species have a scientific name and common name. Scientific names are usually of Latin origin and consist of an italicized capitalized genus and a lowercase species, as established by Linnaeus's binomial nomenclature system. For instance, the scientific name for Xshap/Western Rattlesnake is *Crotalus oreganus*— the name for the genus is the Latin for "rattle," and the species name comes from a nineteenth-century biologist adding the Latin ending -*us* to his misspelling of Oregon, where he found the snake. In this book, I follow the convention of capitalizing common names only when used as proper nouns (e.g., to refer to that specific species and not to those types of snakes as a whole). Subspecies are groups of snakes within a single species that appear to show certain differences from one another and typically have their own common names. For simplicity, I largely avoid discussion of subspecies unless they are threatened with extinction. Readers who want to learn more about Californian snake subspecies can go to www. CaliforniaHerps.com, and those seeking taxonomic information for reptiles of the world can find it at www.Reptile-Database.org.

How to Find and Watch Snakes in California

Finding snakes is a combination of persistence, luck, and skill. Persistence and luck are up to you and the herping gods, but I can teach you a thing or two about skill. The key to skill is knowledge, and this book contains a wealth of knowledge about the habits of California snakes based on twenty-five years of personally herping this beautiful state, dozens of scientific studies on the natural history of snakes in our state, and anecdotes from fellow herpers. You can maximize your herping experience by learning where to go and when to go there. Start small, perhaps by looking for garter snakes

Jeff Lemm

A young herper photographs a Western Rattlesnake from a safe distance.

at your local wetland. If you're visiting California, ask a naturalist or ranger at a state or national park about spots to observe snakes. Once you've got that down, it is up to you to be persistent.

When you find snakes, I strongly encourage you to post photos to the website and app iNaturalist, where experts can confirm your species identification and where you can later browse observations from other herpers. This is especially important for rare species or non-native species because your entries can help scientists study their distributions. However, it doesn't hurt to post common species, too, as all records contribute to a greater understanding of snake diversity in California.

While this book will give you key insights to help you become a skillful herper, I am not going to tell you specifically where to go herping. Rather, I aim to help you find herping spots on your own. I suggest that you avoid asking others to tell you their herping

Looking in crevices is a great way to search for snakes like this Western Rattlesnake.
Photograph by Marisa Ishimatsu.

spots, too, because herpers never share their spots. Is this "herping gate-keeping"? Maybe so. Some people who worked for years to find their own herping spots think others should have to do that as well. But it is also because too many herpers have been "burned" by people with whom they have shared their herping spots. They might return to find the habitat destroyed by someone using a crowbar to get snakes out of rocks or flipping cover objects and not putting them back (both are huge no-no's, as you will learn below). You're going to need to find your own spots, and I'm here to help you learn how to find snakes in those spots.

Here is a list of techniques and tips for finding snakes:

1. Hiking

The single best way to find most snakes is to hike around in appropriate habitats and look for them. Get off the beaten path, where you are more likely to find snakes. Read the species accounts in this book to help figure out where to look for snakes, what time of day to go hiking, and then . . . get out there! Here is a fun tip: Bring a small mirror or cell phone to use to reflect the sun's rays into burrows or rock crevices to look for coils within.

2. Flipping

This is the term that herpers use to describe looking under cover objects for snakes hiding beneath them. Snakes love to hide, and a little bit of knowledge about how to flip can go a long way. Cover objects include rocks, logs, stacks of firewood, pieces of tin, plywood, discarded carpet, cardboard, and basically anything else that provides a snug hiding place for snakes. A herper's playground is a huge junkpile with lots of flippable objects spread out across the ground. Never put your fingers under an object when you flip it, as there could be a venomous rattlesnake hiding underneath it. If you find an animal hiding under the cover object, you *must* move the creature out of the way before putting the object back down, to prevent squishing it. The best way to do this is to pick up or scoot the critter out of the way, replace the object snugly, then release or scoot the animal right next to the object so it crawls back under. Most objects leave an outline in the grass or dirt, so you can fit it back perfectly. It is extremely important that you put the object back exactly as it was found so that the snake's home is restored. This is respectful to the animals and to the herpers who come after you.

3. Driving

Road cruising (a.k.a. night driving, night riding) is a wonderful way to look for snakes because you cover so much ground in a vehicle that you increase your chances of encountering a snake crossing the road. Although it varies by snake species, the best time to road

Brittany App

The author finds a rattlesnake by flipping a cover object in a junkpile.

cruise is at dusk and for the next few hours after sunset because that is when snake activity usually peaks. If you're looking for the kinds of snakes that are active in the morning or during the heat of the day, switch up your timing. Despite the huge variation in color and pattern of California snakes, most of them appear white against the background of a dark paved road at night. Choose roads that cross excellent habitat but have as little traffic as possible. If you're a beginner, you can try driving around 20-25 mph while looking for snakes on the road. As you get better at spotting them, increase your speed to about 35–40 mph to maximize area covered. Snakes are more camouflaged on dirt roads, so drive slower to better spot snakes. When road cruising at night, use your high beams, but be

Road cruising is a great way to search for snakes like this Gopher Snake. *Photograph by Marisa Ishimatsu.*

sure to be courteous and turn these off for oncoming traffic. If you find a snake in the road, always pull over onto the road shoulder and put on your hazard lights, close your doors so they don't extend out into traffic, and be extremely careful when walking around in the road. It is good practice to carefully catch the snake and release it about thirty feet off the road in the direction it was heading.

4. Other methods

Other methods for finding snakes are mainly for use by scientists. It is possible to trap snakes using various contraptions, for example, pitfall traps where a snake falls into a bucket buried into the ground and is later retrieved by a scientist. However, trapping snakes requires approval and special permits from the California Department of Fish and Wildlife.

Think Before You Touch

I first fell in love with snakes as an undergraduate taking a field biology class where we alternated birdwatching, trapping rodents, analyzing predator scat, wading around in water searching for amphibians, and looking under logs and rocks for snakes. The herps grabbed my heart because I could capture them, hold them in my hands, touch their skin, and marvel over their anatomy up close. Snakes were especially desirable because they were relatively few and far between, making the thrill of finding one even more intense.

I wasn't the only one. I have watched generations of my own students at Cal Poly fall similarly in love with the natural world by virtue of a snake in-hand. A recent study examining the field notebooks of introductory biology students confirmed it: Reptiles and amphibians get top mentions among all other organisms when students were asked to reflect on what made them feel connected to organisms.

So, I totally understand why you might want to catch and handle snakes. It's fun! Also, some snakes are really fast, and the only way to really see them up close is to handle them. Taking good photographs of snakes can sometimes require placing them in a spot with proper lighting and surroundings (and lots of patience). Snakes are often found lying across the road, so picking them up and releasing them off the shoulder can save them from becoming road jerky. All of these can be legitimate reasons to handle snakes. However, there are several questions you should ask yourself before catching a wild snake:

1. Is it venomous?

In California, the only snakes that are dangerously venomous to people are rattlesnakes. While rattlesnakes can be easily distinguished from harmless snakes (see page 19), it is possible for

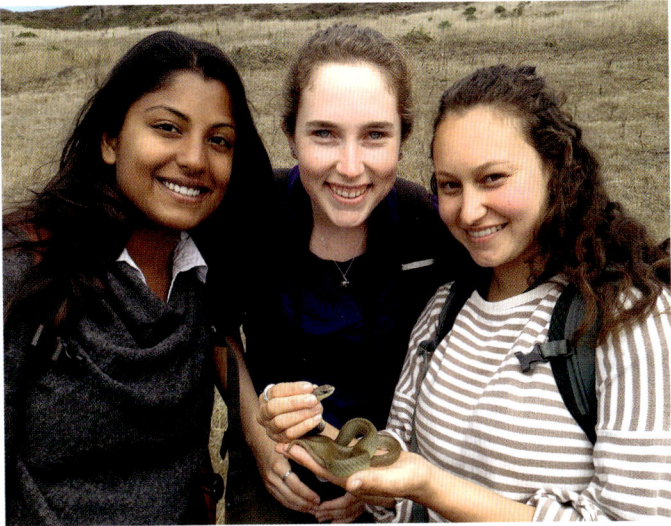

John Perrine

North American Racers and other snakes are popular among biology students.

beginner herpers to mistake a rattlesnake for a harmless snake like a Gopher Snake. Given that the consequences of picking up a snake include potential envenomation, make sure that you are 100 percent positive that the snake is harmless before you pick it up.

2. Do I need a permit?

In California, you must possess a valid fishing license from the California Department of Fish and Wildlife to handle or collect snakes. I know—snakes are not fish! But the fishing license covers several animal groups, including amphibians and reptiles. The license permits you to catch and release any "open season" species of amphibians and reptiles, so long as you are not in a park where handling wildlife may be prohibited. Open season species are those that are not listed as protected (endangered, threatened, or a species of species concern; this latter category is given to species that

Rattlesnakes should be left alone.

may become threatened or endangered in the future if care is not taken to preserve them and their habitat). Protected species must not be handled under any circumstances. California's protected species include those snakes threatened within state limits, even if they have robust populations on the other side of state lines. Notably, a fishing license also allows you to collect and keep up to a certain number of many herp species in captivity (usually two). However, wild snakes should stay in the wild. If you want a pet snake, consider purchasing one from a reputable breeder or adopting a snake that is no longer wanted by its owner (an all-too-common situation given that snakes can live for fifty years or more). For more information on rules relating to wild snakes in California, see the California Department of Fish and Wildlife's regulations online.

3. Will I stress this snake out by handling it?

The answer here is invariably yes. However, briefly handling a snake may only be a minor stressor. If you have a permit and want to capture and handle wild snakes, you should follow a few basic rules to reduce potential stress. First, if you catch a snake, you should release it where you found it within a short period of time, preferably just minutes. Second, be sure that you are not injuring the snake or damaging its habitat by catching it. I have seen herpers use crowbars to pry apart rocks to retrieve Mountain Kingsnakes. This is completely ill advised, and in some areas, illegal. Third, be sure that you wash your hands and any bags or other containers you use in between each snake to limit the spread of snake fungal disease, an emerging pathogen that was found in California for the first time in 2019 and could be spreading throughout the state.

Sure, catching and holding a snake is fun, and you sometimes need to catch a snake to properly identify it. But many times, watching snakes from a distance is even better than catching them. You might see something really cool that you would have missed if you had grabbed the snake straightaway. In some of the species accounts in this book, you will read about incredible experiences I have had with snakes that occurred because I stood back and watched the snake instead of disturbing it. So, think twice about catching every snake you see in the wild! You might be thankful you left it alone.

California Rattlesnakes Deserve Our Respect, Not Fear

Rattlesnakes are as much a symbol of the West as are cowboy boots and cattle. They have been in California far longer, however, and their natural history is tightly woven into the fabric of the

Western Rattlesnake. *Photograph by Spencer Riffle.*

Golden State. Rattlesnakes live nearly everywhere in California, from coastal beaches and tidepools, into forests and grasslands, and across our vast eastern deserts, in some places in incredibly high densities.

Rattlesnakes are highly venomous and arguably even the most dangerous wildlife in California. Most people have never seen a wild rattlesnake and instead have learned to fear or hate rattlesnakes from their unfair portrayal by the media. Many wildlife television programs and YouTube channels show highly defensive rattlesnakes striking at the camera, rearing up, rattling, and otherwise acting terrifying. Like most everything on the television and internet, this is just for views and clicks. In reality, rattlesnakes are gentle creatures that just want to be left alone to go about their snaky business.

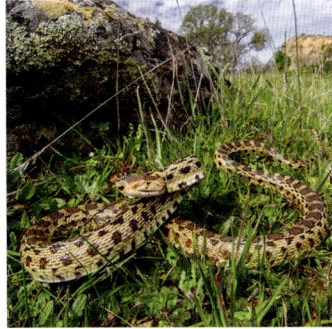

Left: This Red Diamond Rattlesnake is in a defensive posture. Note the rattle on its tail. **Right:** This Gopher Snake may be mistaken for a rattlesnake due to its defensive posture and widened head. However, note its pointy tail, indicating it is not a rattlesnake.

Rattlesnakes only bite people to defend themselves, preferring to use their venom to dispatch and tenderize the prey that they will eat whole. Still, accidents do happen, and the consequences can be very severe. A little bit of knowledge goes a very long way in keeping people and their pets safe from rattlesnakes (and likewise in keeping rattlesnakes safe from people and their pets). Follow these guidelines to be respectful of rattlesnakes when you're out enjoying nature, and you need not be fearful about these amazing animals.

TIP #1
Learn to distinguish rattlesnakes from harmless snakes.
- Learning to identify snakes requires using multiple characteristics and lots of practice. Use the species accounts in this book to hone your identification skills over time.
- Rattlesnakes have an appendage on the tips of their tails made of keratin (like your fingernails), and each segment of this rattle loosely interlocks with the next segment such that it makes a high-pitched buzzing noise when the snake shakes its tail in warning.

- Notably, rattlesnakes do not always rattle in warning, and many harmless snakes will shake their tails when they are nervous.
- Baby rattlesnakes cannot make noise with their rattle until they shed (usually during their second year), but they do have a rounded "button" on their tail.
- If you find a snake tightly curled up like a cinnamon roll, it is likely a rattlesnake. Other snakes seldom do this.
- Isn't it true that rattlesnakes have wide heads and skinny necks, vertical pupils, plus rough-looking and dull skin? While these characteristics are *often* true of rattlesnakes, they are not reliable ways to distinguish rattlesnakes from other snakes. Don't touch a snake unless you are 100 percent sure that it is not a rattlesnake.

TIP #2

Prevent accidental snakebite while hiking by following these guidelines:

- Watch where you put your hands and feet when engaging in outdoor activities.
- Stay on trails and avoid tall grass when in the outdoors.
- If you need to work in tall grass, wear sturdy boots that cover your ankles and preferably also your calves. Avoid footwear with breathable mesh panels, as these do not block a rattlesnake's fangs. Snake gaiters can also be worn to protect your calves.
- Keep your dogs on leash while on hikes. For dogs that run free (like hunting dogs) or if rattlesnakes enter your yard, consider enrolling your dog in a rattlesnake avoidance training workshop. Always keep cats indoors, as they never belong outdoors.
- If you see a rattlesnake in the wild, admire it from a distance and give it a wide berth (Ten feet is perfect.).

Always carry a phone in case of emergency.

- From most areas, you can call 911 from your cell phone in case of a snakebite emergency.
- If you frequently hike or camp in areas with no cell reception, you should have a satellite phone or GPS messenger device.

If you live in a rural area or near open space, reduce the chances of rattlesnakes hanging around in your yard by following these guidelines (Of course, if you want to attract rattlesnakes to your yard, simply do the opposite of these!):

- Replace full, bushy landscaping with "spindly" plants that do not offer hiding places.
- Patch any openings leading under or into your home or outbuildings so that rattlesnakes cannot hide in them.
- When cleaning up debris or gathering firewood, never put your hands under objects without looking for snakes that could be hiding underneath.
- Avoid bird feeders, as these can attract rodents that then attract snakes.
- Reduce use of water and avoid water sources (leaky faucets, fountains, pools, irrigation) because they attract snakes, especially during prolonged droughts.
- If your yard is regularly visited by rattlesnakes, consider investing in rattlesnake exclusion fencing to permanently keep them out.

What Good Are Rattlesnakes?

1. Rattlesnakes eat huge numbers of rodents, helping to keep populations of these herbivores at appropriate levels. This protects native plants, our crops, and helps to control numerous diseases that can be spread by rodents, including Lyme disease, hantavirus, and bubonic plague, all of which occur in California.

2. Rattlesnakes are extremely efficient in converting their meals into snake biomass, then in turn they act as essential prey for our beloved mountain lions, bobcats, coyotes, badgers, hawks, and owls. So, rattlesnakes provide protein and fat-packed snake-sausage dinners to our top predators.

3. Rattlesnakes even play a role in dispersing seeds! When a rattlesnake eats a rodent with cheek pouches full of seeds, the snake's gut digests the rodent but not the seeds. Then a couple of weeks later the snake poops the seeds out, usually far away from the parent plant. So, rattlesnakes rescue and disperse the seeds that would have been digested by the rodent, even providing the seeds with mineral-rich fertilizer (poop) in the bargain.

4. Rattlesnake venoms are a source of inspiration for drug development. There are already several drugs derived from rattlesnake venom on the market (most of them harness the anticoagulant activity of venom to help prevent blood clots), and others are in development.

In case of snakebite to a person or pet:
- Remove all jewelry from the limb or pet collars in case of swelling.
- Do not do anything to the wound (no ice, no heat, no pressure, no tourniquet, no cutting/sucking, et cetera).
- Go straight to the nearest hospital or emergency veterinarian. Calling ahead to make sure they carry antivenom is a good idea. Antivenom is the only treatment that can successfully treat a rattlesnake envenomation in California. Avoid steroids, antihistamines, prophylactic antibiotics, and non-steroidal anti-inflammatory pain medications, as these are all either useless or potentially harmful in snakebite victims.

Avoid handling rattlesnakes unless you have received expert training.
- If there is a rattlesnake in your yard, you might be able to scare it away by spraying a hose at it. In many areas, you can also call a paid or volunteer service to relocate the snake. You can try calling the local fire department or animal control, but in many areas, they are obligated to kill the snake.
- Do not pick up or touch rattlesnakes. Those YouTubers that free-handle rattlesnakes? They are playing with fire, setting a bad example for youngsters, and being disrespectful to the snakes. Don't touch dead rattlesnakes, either. A rattlesnake that has been hit by a car may still be able to bite. Snakes that have been beheaded can indeed still bite for hours afterward. Also, don't behead rattlesnakes. It is dangerous and cruel.
- Many rattlesnake bites are to the hands of young males, often with alcohol involved. Don't drink and handle snakes. Enough said.

Now that you know how to be safe when enjoying the great outdoors in rattlesnake country, I hope that you will head outside with renewed confidence. If you are lucky enough to see a rattlesnake in the wild, give it plenty of space and watch it for a while, and think about what it is up to. A coiled-up rattlesnake might be sitting in ambush position waiting for an unsuspecting mouse to run by. Or, if it has a big food bulge, it may have already eaten and is sunning itself to help it digest. A rattlesnake on the move has places to be—could the big snake you saw crawling across a trail have been a male following a potential mate's scent trail? If you encounter multiple rattlesnakes together in the summer, you might have found a rattlesnake rookery, or a communal nest where females give birth and help each other care for their pups. Whatever you find, I hope that you will share photos with your friends and families to help spread the good word about these fascinating creatures.

That's Not a Snake!

When people see a critter that lacks legs—or *appears to* lack legs—they often jump to the conclusion that it is a snake. Because so many people are afraid of snakes, I get lots of phone calls from people worried about the "snake" in their garden or garage. I always ask for photos so I can help them figure out what to do. Here are a few animals that are often confused with snakes.

In California, legless lizards of the genus *Anniella* are commonly misidentified as snakes. However, they are actually lizards that are closely related to alligator lizards! If you have good enough eyesight to inspect them close up, you will see that they have eyelids—something that snakes never have.

Several other species of lizards and salamanders are sometimes confused with snakes because of their elongated bodies and tiny limbs. Other critters sometimes mistaken for snakes include millipedes, centipedes, and some caterpillars.

Marisa Ishimatsu
Marisa Ishimatsu
Marisa Ishimatsu
Jeff Lemm
Marisa Ishimatsu

These animals are commonly mistaken for snakes. Clockwise from top: San Diegan legless lizard (genus *Anniella*), skink (lizard in genus *Plestiodon*), slender salamander (genus *Batrachoseps*), whiptail (lizard in genus *Aspidoscelis*), alligator lizard (genus *Elgaria*).

THE SNAKES

Spencer Riffle

RUBBER BOA

CHARINA BOTTAE

FAMILY BOIDAE

Rubber Boas are generally more common in the northern parts of California, so those of us farther south need to travel northward to get our boa fix. I encountered my first Rubber Boas in the Bay Area over a decade ago, when some colleagues invited me to visit their field site in coastal San Mateo County where they were doing herpetological surveys. They had a massive array of cover boards—literally hundreds—and we spent all morning hiking through the beautiful coastal grassland, checking under the boards to see what reptiles were sheltering there. It was herp heaven. Most of the boards yielded at least one critter, sometimes more. Second only to the ridiculously abundant North American Racers (see page 48) were the Rubber Boas, which we found by the dozens. This field trip spoiled me, because it turns out that Rubber Boas are not always so easy to find. I now live in San Luis Obispo County, on the California Central Coast, where Rubber Boas only occur in an isolated population in Montaña de Oro State Park. My students and I have spent countless hours hunting for this snake with no luck. In the past fifteen years, just a handful of Rubber Boas have been confirmed in the park, most from photos sent in by visitors. Finally, while I was writing this book, a friend found a boa that had been run over by a mountain bike on a trail in the park, and we deposited the specimen in a research museum where it can be studied. Still, it goes to show you how spotty their distribution and abundance is in California.

Appearance: Rubber Boas are small snakes with brownish-yellow to grayish skin that often sags around the snake's body, and their undersides vary in color but are often yellow or pinkish. They have small, granular scales with larger scales on their heads and bellies. Their small eyes with vertical pupils give them a comical, goofy look. They are "pencil necks," meaning their heads are the same width as their bodies, and their snouts taper to a rounded point. Another defining characteristic of this species is that their tail tips are rounded instead of pointy, sometimes making it hard to tell which end is the head and which is the tail.

Natural History: Although Rubber Boas can be found in a variety of habitats, they are most common in grassland and forest habitats, and are typically found in moist, riparian corridors and rocky areas. They spend a lot of time burrowing underground or resting under logs and other cover objects, but you might also get lucky and see one out crawling in the open. They are dietary generalists, eating basically any small vertebrate they can capture by grabbing them with their jaws and constricting them. Rubber Boas are thought to mate in spring shortly after emergence from winter dormancy, and the live-bearing females give birth in the late summer or fall.

Range and Variations: Rubber Boas have been recognized as either a single species or two, occupying coastal areas from San Luis Obispo County northward into Canada, as well as in the Sierra Nevada and Great Basin eastward to Colorado, Utah, and Montana, plus the San Bernardino and San Jacinto Mountains. A recent study found that there are multiple genetic lineages in California, but for now the scientists advocated recognizing a southern lineage as a subspecies, the Southern Rubber Boa. The Southern Rubber Boa is much smaller in size than northern individuals and may eventually be protected by the California Department of Fish and Wildlife as their habitat disappears due to development and climate change.

How to Find Rubber Boas: As I know all too well from struggling to catch one near my home, Rubber Boas can be hard to find! They occur from sea level up to about 9,000 feet elevation, and they prefer low temperatures. Your best bet is to avoid searching for them when it is hot and dry, and instead focus on herping for them when it is cooler and wetter outside. They may crawl around in the open at dusk, nighttime, or dawn, but the best way to find them is to look under cover objects like boards, logs, and rocks in appropriate habitat. In high elevation, inland areas of California, you can also night drive for Rubber Boas in the summer.

Marisa Ishimatsu

You Might Like to Know: Rubber Boas have an incredibly endearing anti-predator response. When hassled by a predator (or a handsy herpetologist), a Rubber Boa curls up and hides its head deep inside its coils while waving its tail around. The first time this happened to me I did a double-take—that stubby tail sure did look like a head! This behavior could save the snake's life by tricking a predator into grabbing the snake's tail instead of its head, possibly allowing the snake time to fight back and escape only with injuries to its tail. Sure enough, Rubber Boas are often found with healed scars on their tails, suggesting that this behavior might really work.

Spencer Riffle

ROSY BOA

LICHANURA ORCUTTI

FAMILY BOIDAE

Ah, the coveted Rosy Boa. This beautiful snake is highly prized among field herpers and captive breeders alike. I have some experience herping for these snakes in the wild, though most of this was when I lived in Arizona in graduate school where we saw boas readily on hikes in its western Sonoran Desert habitat. In California, a parallel might be the hills in San Diego County, where Rosy Boas can be found aplenty if you know where to look. This snake is a perfect example of a species that you need to learn how to hunt by putting in the time hiking in good habitat, as few other herpers will share their spots for fear that someone will poach or otherwise harm the snakes. They are worth all the hours spent wandering the desert just to see them in their natural environment.

Appearance: Rosy Boas are one of the most beautiful snakes in California, and indeed in the world. They are also very distinctive, and it is unlikely that you will mistake them for any other species. They are medium in size (usually 2–3 feet in length), stout-bodied, and most often their pattern consists of several grayish-white and dark orangish-pink stripes that run the length of their bodies. Their scales are shiny and reflective, and their eyes are rather small. Their tail is somewhat pointier than that of a Rubber Boa, but still more blunt than that of other snakes.

Natural History: Rosy Boas are more often found in low and mid-elevation rocky desert habitats, especially areas with large boulders. It is rare to find Rosy Boas active during the day; instead, you are most likely to find them crawling on the desert floor from dusk until dawn. They readily eat most any type of small verte-brate, killing them by constriction. Rosy Boas mate in the spring, and the females give live birth in the fall.

Range and Variations: Rosy Boa taxonomy has changed a lot over the years. Until recently, the species was named *Lichanura* (or *Charina*) *trivirgata*, with two subspecies occurring in California, one in the inland desert and one near the coast in extreme Southern California. Some scientists recognize the southern form as its own species, but for this book I group them together into a single species due to the controversial nature of the taxonomy. Adding to the controversy is the notion held by some that the scientific name

Zeev Nitzan Ginsburg

of this species should actually be *L. roseofusca.* More interestingly than names, in my opinion, is a population of "unicolor" Rosy Boas in San Diego County that lack stripes and are a beautiful, solid coppery color.

How to Find Rosy Boas: Rosy Boas are a great example of a snake where the strategy for finding them depends on the habitat. In inland deserts, your best bet is searching in rocky areas at 3,000–5,000 feet elevation, while closer to the coast they can be found lower in elevation, too. In cool weather, search in rocky crevices or under rocks. In warm weather, go hiking with a strong flashlight shortly after sunset, or night drive for Rosy Boas on remote roads that transect rocky areas.

You Might Like to Know: Compared to most other snakes, relatively few people have found Rosy Boas in the wild. They are nonetheless one of the most iconic Californian snakes. This is largely because they are hugely popular in the pet trade. Most of what we know about their reproduction was learned via captive breeding. Families regularly ask me what kind of snake to get for their eager, snake-loving children, and I always suggest that they choose from a Corn Snake, a Ball Python, and a Rosy Boa. Rosy Boas tend to be gentle, eager to eat, and easy-to-care for snakes that make excellent first pets. They mate readily in captivity, and breeders have coaxed multiple beautiful morphs into existence by selecting for certain colors and patterns. If you want a pet Rosy Boa, be sure to buy one at a reptile expo instead of collecting a wild snake.

Spencer Riffle

GLOSSY SNAKE

ARIZONA ELEGANS

FAMILY COLUBRIDAE

Recalling my first encounter with a live Glossy Snake always makes me laugh. Though I learned about them in my herpetology class in college in California, I didn't see one in person until I began grad school in Arizona. My advisor and I regularly went night driving, and when the summer monsoons began, we would speed around the desert backroads, gluttonously admiring snakes of all shapes and sizes. One night, I jumped out of the car to get a snake, picked it up, and announced it to be a Gopher Snake. My advisor peered at it, began cracking up, and scoffed to me that it was in fact a Glossy Snake, and *hooooowwwwwww* did I not know how to distinguish these two species?? Oops. In my defense, they do look *kind of* similar. It just goes to show that we are all beginners until practice makes perfect. Now I regularly see Glossy Snakes while night driving in the desert, and I am proud to say that my students are better beginner herpetologists than I was: When I ask them to identify this beautiful, shiny snake, they joyously yell out in unison, "Glossy Snake!"

Appearance: As the name implies, Glossy Snakes are very shiny and smooth in appearance. These medium-sized snakes typically have a mottled pattern with narrow brown or grey saddles on the top and brown spots on the sides, on a tan or yellowish-pink background. Overall, in California they tend to be very light in color, more so than Gopher Snakes. Their faces are gently sloped into a somewhat pointed nose, and they usually have a dark stripe going from their cheeks through each eye and meeting in the middle in an adorable "unibrow."

Natural History: Glossy Snakes are found primarily in deserts and grasslands, where they hide in burrows or under the sand during the day and can be found hunting above ground at night. Although they have a varied diet that includes rodents, lizards, and snakes, the main way that they hunt is by sniffing out lizards sleeping at night, then either constricting them or just wolfing them down alive. Mating occurs in the spring, females lay eggs in the early summer, and hatchlings emerge later in the summer.

Marisa Ishimatsu

Range and Variations: Glossy Snakes range eastward into Texas, south into northern Mexico, and northward as far as Nebraska. In California, they are mostly a Southern California species.

How to Find Glossy Snakes: Glossy Snakes are very much a nocturnal species and are easy to find by night driving through desert or grassland habitat. Glossy Snakes are the perfect example of a snake species that is super abundant in some spots and is extremely rare in others. For example, I've found a great spot in the Mojave National Preserve where, year after year, my students and I find them while road cruising. On this road, I've observed that

Glossy Snakes are common only at higher elevations and come out when it is cool enough that other snakes have gone to bed. In some other areas, this rule doesn't apply at all.

You Might Like to Know: Glossy Snakes were the subject of early studies back in the 1960s and 1970s on the role of the pituitary gland in snakes (Spoiler: it releases hormones that impact shedding frequency in snakes!). Also, the California Glossy Snake subspecies is one of a handful of Californian snakes chosen to have their genome sequenced as part of the California Conservation Genomics Project because it was selected as a "high priority" snake due to its decline in coastal Southern California. So, if you see one of these snakes in the wild, be sure to report it on iNaturalist.

Mike Pingleton

BAJA CALIFORNIA RATSNAKE
BOGERTOPHIS ROSALIAE

FAMILY COLUBRIDAE

The Baja California Ratsnake is in the running for the most beautiful snake in California, and it definitely wins the crown for the rarest snake. As its name implies, it is endemic to Baja California, Mexico, and its range extends just *barely* into the United States in extreme southern Imperial County. Like pretty much everyone else (see below), I haven't seen a ratsnake in California. However, at least one specimen has been found in California, and another could be found in the future. I have found them in Baja California, Mexico, a true herper's mecca. I recall one particularly gorgeous ratsnake I found while night driving in a riverbed outside of La Paz in southern Baja about fifteen years ago. When the headlights hit that long, sleek, pink body, there was no doubt about what it was. Shrieks of "Mira la culebra!" and "OMG!" poured out of the vehicle as we literally jumped through the car windows to see this gorgeous snake up close and personal. It was pushing five feet in length and was fat and healthy. Ah, to encounter this mysterious snake species again!

Appearance: Baja California Ratsnakes are long, thin snakes, potentially reaching up to 5 feet but more typically 3–4 feet. They have long, wide, flat heads with big eyes and squared-off noses. Adult snakes are a uniform light-brown color, often with a pinkish, yellowish, or olive hue, that stands out against the darker skin that is visible between scales. The belly scales are much lighter in color. Hatchlings have a vague blotchy pattern of darker brown on a slightly lighter background, but this disappears with age.

Natural History: Baja California Ratsnakes are most often found in rocky lowland desert habitats, sometimes in the vicinity of dry or flowing creeks or other riparian areas. Few studies have been done on this species, so their habits are largely unknown. They likely spend much of the day hiding underneath rocks or in rocky crevices, then become active mainly at night. Adult ratsnakes actively forage for small rodents and kill them by constriction. Mating likely occurs in the spring, shortly after which the females lay eggs, and then hatchlings emerge in the summer.

Range and Variations: Ranging primarily in Baja California, the Baja California Ratsnake has been confirmed in California by only one specimen, found dead on the road. The best herpers have

Marisa Ishimatsu

scoured nearby areas in search of this species and have come up empty-handed. That said, the species can be found only a few miles south of the border in similar habitat, suggesting that the Baja California Ratsnake could indeed exist in California.

How to Find Baja California Ratsnakes: If you really want to see a Baja California Ratsnake, you should search for one in Baja California, not in California. Because these snakes are so spotty and rare even in many places in Mexico, you will maximize your chances by night driving though rocky areas, although you certainly could get lucky by hiking at night with a good flashlight or even in the daytime near areas with water, like desert springs and oases. If you do find a ratsnake in California, it is essential that you document this find with photographs and an accurate GPS locality, then report this observation on iNaturalist.

You Might Like to Know: I would love to tell you all sorts of juicy details about this large, gorgeous snake. But the truth is that we know very little about it. In fact, it is probably the least studied of all California snakes. This isn't just because it barely enters California—even in Mexico, it has not been studied much at all. Reasons for this are complicated. In general, there are fewer resources for biological study in Mexico than in the United States, though this is changing rapidly as numerous Mexican herpetologists are entering the scene. The paucity of research on this snake could also reflect that it tends to be relatively scarce and rarely observed in the wild. I can only hope that in future editions of this book, I will be able to update this section with news from studies by young scientists inspired to learn about the natural history of this beautiful snake.

Jeff Lemm

SHOVEL-NOSED SNAKES

CHIONACTIS ANNULATA AND *C. OCCIPITALIS*

On my very first herping trip to the Mojave Desert back in the '90s, I was walking through a lava flow in the middle of the day and saw a colorful squiggle in my peripheral vision start to move. When I turned to look at it, I saw a little snake with bright yellow, red, and black bands trying to escape down into the sand. I snatched this curious little snake up before it could get away, used my field guide to key out the snake, and was delighted to identify it as a Mojave Shovel-nosed Snake. These snakes are usually out at night, but I found it during the day! The characteristic that gave it away was its pronounced "overbite," where its lower jaw is recessed compared to its upper jaw. This helps to give these little snakes a streamlined head shape, allowing them to burrow easily through loose sand. When I let it go after admiring it, it whooshed right into the sand. If you ever find a shovel-nosed snake, for example, while rescuing one from the road, I highly recommend letting it go in a sandy patch near where you caught it so you can watch it sand-swim away from you.

Appearance: Shovel-nosed snakes are very small, usually 1 to 1.5 feet long and only about as big around as a pencil. The lower jaw is deeply sunken. Their scales are smooth and shiny. Shovel-nosed snakes always have dark bands on their backs that usually end on their sides but sometimes encircle the whole body. The background color is cream or yellowish and, sometimes but not always, they have red bands too. The Colorado Desert Shovel-nosed Snake is more likely to have dark bands that completely encircle the body and to have red bands than the Mojave Desert species.

Natural History: Shovel-nosed snakes inhabit sandy areas in the desert, where they bury themselves under the sand until they emerge at night. They feed primarily on invertebrates, sometimes while swimming under the sand and sometimes while out on the surface. In much of their range, the bulk of their above-ground activity occurs in late spring and early summer, when mating occurs. Females lay eggs shortly thereafter, and these hatch in the summer.

Range and Variations: There are currently two species of shovel-nosed snakes recognized in California: the Mojave Shovel-nosed Snake (*Chionactis occipitalis*), which can be found throughout

Mojave Shovel-nosed Snake. *Photograph by Joshua Wallace.*

the Mojave Desert in Eastern California as well as Nevada and Arizona, and the Colorado Desert Shovel-nosed Snake (*C. annulata*), which can be found in the deserts of Southeastern California. Some scientists believe that shovel-nosed snakes should be lumped into the same genus as ground snakes (*Sonora*); these species indeed share numerous characteristics and can be tricky to tell apart.

How to Find Shovel-nosed Snakes: I recommend night driving on desert roads that transect areas with loose sand to find shovel-nosed snakes. In some parts of the California deserts, these species can be extremely common and you might see many in a single night, especially from late April through June. As the hot summer progresses, they usually spend more time underground in the sand that they are so well adapted to live in. They even shed down in the sand! However, they always come above ground to poop, for reasons unknown to science (tell me if you have a hypothesis!).

You Might Like to Know: The diminutive size and derpy-looking overbite make these little snakes rather endearing. But if you were a scorpion, a shovel-nosed snake might be the stuff of nightmares to you! When I watched the recent film adaptation of the science fiction novel *Dune*, the sand-swimming predatory "worms" reminded me a lot of shovel-nosed snakes. They swim easily through the sand, have valved nostrils to prevent sand from getting up their noses, and can burst through the sand underneath an unsuspecting scorpion and gobble it right up before diving back down into the sand. They can withstand multiple stings from scorpions with no apparent ill effect, suggesting that they are resistant to scorpion venom.

Ryan Sikola

NORTH AMERICAN RACER

COLUBER CONSTRICTOR

FAMILY COLUBRIDAE

In the afternoons, when they are warmed up and out stalking lizards, these snakes make it easy to understand why they are called racers! But finding them can sometimes be tricky. The North American Racer is the perfect example of a snake whose abundance is what I call "spotty." In many areas, you might see one now and then. But in some places, their numbers are truly astounding. When I visited a big stretch of habitat north of Santa Cruz in California with some friends who were monitoring a series of cover boards as part of a herp survey, I encountered super high numbers of racers. I had seen one or two of these on field trips in college in the Bay Area, and I have seen a handful of them on the central coast when out with my own students, but nothing compares to that board line. There were racers under almost every board. Sometimes two, sometimes more. And they weren't "racery." Instead, they were cold and sluggish, graciously allowing us to take their measurements and release them back under their cover boards to continue sleeping the cold morning away.

Appearance: This medium-sized snake is typically shiny olive green or brown on top with a bright yellow belly, though sometimes the belly can be a bit whitish. These snakes are notable for having big eyes, which give them away as being active during the day. Their long, thin body shape facilitates their zoomy lifestyle; often you only see a green flash disappearing into the brush. The hatchlings have spots on their backs that make them look like a completely difference species.

Natural History: The prime habitat for North American Racers is grassland, where they blend in with the greenish-gold grasses of California and are camouflaged from prey and predators alike. They can be found in chaparral and woodland, too, but typically in open areas with plenty of sunlight shining through. Based on the scientific name, which means snake (*Coluber*) that constricts (*constrictor*), one would be forgiven for assuming that these snakes constrict their prey, but they do not! They stalk small vertebrates and insects by sticking their heads up like a submarine's periscope, then grab them with their jaws and force them down the hatch, sometimes while still alive and struggling. Mating occurs in the summer, though in some areas it probably occurs in spring too. Females lay eggs in the summer, with the unique-looking hatchlings appearing on the scene in the late summer and early fall.

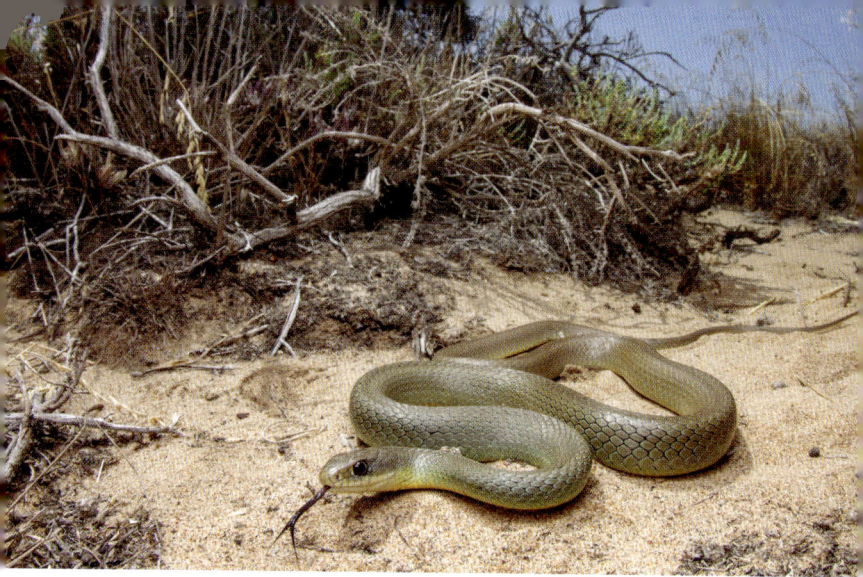

Francesca Heras

Range and Variations: The North American Racer is extremely widespread, ranging across the entire United States down into Central America. The only subspecies in California is called the Western Yellow-bellied Racer, a name I include here because it is used more often than the official name of North American Racer and is found throughout the American West and into Southern Canada. In California it can be found most anywhere outside of the deserts and southern Central Valley.

How to Find North American Racers: As you can tell by the introduction to this species, cover boards can be extremely effective in luring North American Racers to take up residence in some parts of their range. So, flipping cover objects, including natural objects like rocks, can be fruitful. Aside from that, hiking on warm days in meadows and other open spaces are your best bet for seeing a racer.

You Might Like to Know: To me, hatchling North American Racers are like pumpkin spice: They mark the coming of fall. On the California Central Coast, racer hatchlings always start to appear around the beginning of September. The hatchlings have spots and look different enough from adult racers to confuse people about their identity. The spotted pattern on juvenile racers is likely an adaptation to help avoid predators. The idea is that the spots may produce a confusing visual effect as they speed away, whereas in adults the benefit of solid green color to blend into the vegetation is more important. But I also think that the spotted pattern of baby North American Racers could be adaptive as a form of mimicry of baby rattlesnakes. Racers hatch at the same time as rattlesnakes are born, and a bird or other predator might think twice when they see that spotted pattern because it could indicate that the meal is "spicier" than expected. This hypothesis remains to be tested experimentally, and what a fun experiment it would be. If any-one wants to join my lab as a student to do this research project, hit me up!

Marisa Ishimotsu

Spencer Riffle

SHARP-TAILED SNAKES

CONTIA LONGICAUDA **AND** *C. TENUIS*

FAMILY COLUBRIDAE

About a year ago, I moved to a lovely, rural, unincorporated area outside of the town of Santa Margarita, in San Luis Obispo County. Unbeknownst to most of its residents, Santa Margarita has the honor of being the southernmost city home to sharp-tailed snakes. Here in Santa Margarita and in many similar towns all the way up to Canada, sharp-tailed snakes are commonly found underneath logs and within the soil of most backyards. I begin and end each day literally surrounded by subterranean sharp-tailed snakes, though for some inexplicable reason they are absent around my workplace a mere ten miles to the south. I am nerdy enough to ponder deep questions about the ways of sharp-tailed snakes on my daily commute. Where exactly is THE southernmost sharp-tailed snake? Why don't they occur further south? What eats all the slugs south of Santa Margarita, if not sharp-tailed snakes? Do slugs have big parties with conga lines because they don't need to worry about these fierce predators? And just think—what questions might your own encounters with sharp-tailed snakes provoke?

Appearance: Sharp-tailed snakes are tiny little snakes with—you guessed it—sharp, pointy tails. They are usually less than 1 foot long, with the biggest about 1.5 feet. They seem smaller, though, because they are so narrow and lightweight. These snakes usually have a greyish or olive background, sometimes with faint red, pink, or orange stripes along the sides of their backs. Their bellies are typically light with black bands across each scale.

Natural History: Sharp-tailed snakes are usually found in forests, in surrounding grasslands, and in people's yards in forested areas. They are primarily burrowers, inhabiting moist soils, and rarely found out on the surface. No one knows much about what they do when they are active underground, but studies of their gut contents show that they eat almost exclusively slugs and slug eggs, and sometimes slender salamanders. Little is known about sharp-tailed snake reproduction except that they lay eggs in summer that hatch in fall. Presumably, the snakes mate in spring. As temperatures rise and the California soils dry out, the snakes curl up deep in the soil until temperatures drop and soil moisture increases.

Range and Variations: Only one species of sharp-tailed snake was designated until about ten years ago, when researchers used DNA analysis to discover that certain populations in coastal Oregon and Northern California are actually a separate species (Forest Sharp-tailed Snake, *Contia longicauda*) from the one in the rest of Northern California (Common Sharp-tailed Snake, *C. tenuis*). It's

Marisa Ishimatsu

Common Sharp-tailed Snake.

not easy to tell these species apart by eye. While there are some slight differences (e.g., Forest Sharp-tailed Snakes have longer tails, hence their name *longicauda*, "long tail"; Common Sharp-tailed Snakes have wider black bands on the belly), the easiest way is based on location. In California, Forest Sharp-tailed Snakes only occur in a narrow strip along the coast in the South Bay (San Mateo, Santa Clara, Santa Cruz Counties) and in Mendocino County north of Del Monte County. Common Sharp-tailed Snakes have a spotty distribution throughout inland and coastal Northern California.

How to Find Sharp-tailed Snakes: To find these snakes, flip cover objects like logs in grassland at the edge of forested habitat. The best time to do this is in spring when the soil is moist, though in some areas in Northern California you can find these snakes under cover objects in the middle of the winter and well into the summer. Sometimes you can also find them in the fall after the first rains. Gardening enthusiasts might be lucky enough to find them while digging in their gardens and often mistake them for earthworms. If you happen to have a large, forested property, you could even put out cover boards (plywood is great, and corrugated metal and pieces of carpet also work well), and sharp-tailed snakes and other reptiles, amphibians, and arthropods, even nesting mammals, will colonize them.

You Might Like to Know: Sharp-tailed snakes were once thought to be rare snakes, but we now know that they are incredibly common. Scientists initially thought they were rare because they are so small and spend most of their lives underground. This highlights how biased science is against small, secretive snakes. We have learned much about the ways of big snakes like rattlesnakes because they are above ground so often and because their heavy bodies allow them to carry radio-transmitters for scientists to study them. However, tiny little burrowing snakes attract very little attention. If sharp-

Forest Sharp-tailed Snake.

Marisa Ishimatsu

tailed snakes were scaled up, their proportionally huge teeth—an adaptation for grabbing slugs—would be very noteworthy indeed. Also, they can be found in huge congregations, with ten or more snakes sometimes found under the same piece of wood. This suggests that sharp-tailed snakes may exhibit social behavior that we cannot even begin to understand. Finally, I know most of you are wondering what the purpose of the sharp, pointy tail is. While I would love to report that they stab predators with it, that does not appear to be the case. They may use it to help anchor their bodies in the soil as they attack prey, but this is merely a hypothesis, and it remains a mystery like most aspects of sharp-tailed snake biology. One person on social media concluded that the purpose of the pointy tail is "to look metal as heck," and I cannot disagree.

Spencer Riffle

RING-NECKED SNAKE

DIADOPHIS PUNCTATUS

FAMILY COLUBRIDAE

During my days as a budding herpetologist in Berkeley, I saw lots of Ring-necked Snakes. They were common under rocks and other cover objects. Sometimes we would see them crawling around on the surface, especially on cool, wet mornings. You had a good chance of finding a Ring-necked Snake on any given day. Fast forward to living on the central coast, as I do now. These snakes are here, of course, but for reasons that only the snakes know, I see them far less frequently than I did in Berkeley. I had taken for granted this most beautiful of snakes because they were common, but now that I live in a place where they are more rarely seen, I rejoice disproportionally when one of my students finds one on a field trip.

Appearance: Ring-necked Snakes are very small serpents with a solid, dark grey color, an orange ring around their necks, and a brightly colored red-orange underside that is stippled in black. There aren't many other snakes with which to confuse the Ring-necked Snake. Some black-headed snakes have a pale ring around their necks, but they lack the bright red-orange coloration typical of Californian Ring-necked Snakes.

Natural History: Ring-necked Snakes spend a lot of time underground or beneath cover including rocks, fallen logs, and artificial cover objects like boards. You can find Ring-necked Snakes in many habitats, including moist riparian areas, grasslands, chaparral, and forests. Ring-necked Snakes grab their prey and kill it with venom injected via two fangs on the rear of the upper jaw. While this venom is harmless to people, it can rapidly kill their main prey items, including salamanders and frogs, small snakes and lizards, and worms and small insects. Reproduction has not been studied much in Californian Ring-necked Snakes, but if they act similarly to more eastern populations of the species, then they likely mate in spring, females lay eggs in the summer, and tiny hatchlings appear in late summer.

Range and Variations: As a species, Ring-necked Snakes have a wide geographic range, from southern Canada to northern Mexico, and throughout the American South, Midwest, and Northeast. In California they are found in coastal areas and in the Sierra Nevada. Their distribution in the West is spotty because they do not occur in lowland deserts.

Marisa Ishimatsu

How to Find Ring-necked Snakes: Flip, flip, flip those cover objects for a chance to find a Ring-necked Snake hiding underneath them! You might also get lucky and find one out crawling around, but your best bet is certainly to flip logs, rocks, tin, boards, and anything else a little snake could shelter underneath. I have had the greatest success flipping for Ring-necked Snakes in the spring when the ground is moist and the amphibians these snakes like to eat are active.

You Might Like to Know: No discussion of Ring-necked Snakes would be complete without acknowledging the charming posture that they adopt when they are defensive. If you are lucky enough to find a Ring-necked Snake under a log, and if you then disturb it, it will immediately curl up its tail into an ornate coil and flash you with its bright orange-red underside. The color red is often used by

Spencer Riffle

animals to signal a warning to would-be predators; in particular, it can often indicate that the prey is poisonous or unpalatable. Some animals ride on the coat tails of these trailblazers by showing off a red warning display even though they are not actually toxic—this is known as mimicry. So, which is the Ring-necked Snake? Well, it definitely isn't toxic to eat. It does have venom that could potentially be effectively used to defend itself against small predators, like another snake that tried to eat it. That said, it appears that the Ring-necked Snake is mainly a mimic, a bluffer that shows off its red belly despite being harmless to a predator. The odor of the musk that these snakes so readily exude as they coil up defensively is possibly bad enough to count as toxic if you have a sense of humor (and of smell). While among the stinkiest of all Californian snakes, these little bluffers more than make up for it in charm.

Specner Riffle

NIGHTSNAKES

HYPSIGLENA CHLOROPHAEA AND *H. OCHRORHYNCHA*

FAMILY COLUBRIDAE

Just look at these snakes. Aren't they stunningly beautiful? From 5–6 feet above ground, the vantage point of a typical human head, you might miss the intricacies of these beautiful little snakes. If you are lucky enough to find a nightsnake in the wild, squat down to examine them up close and personal in their world, or better yet, get on your belly in the dirt and look at this little serpent head on. You will be amazed at their complex patterns, their distinguished dark collars, and their gorgeous, shimmering eyes. These snakes are seldom seen by people who aren't actually looking for snakes, and it's a darn shame because they are more handsome than just about any other snake when you get close enough to really see them.

Appearance: Nightsnakes are small, shiny snakes that are generally tan or grey in color with a pattern of regular brown blotches on their backs, a set of prominent dark blotches on either side of the neck that look like an ornate collar, and golden eyes with narrow, vertical pupils. They are usually not much larger than a pencil, and many people mistake the tiny full-grown adults for baby snakes.

Natural History: Nightsnakes are found in many different habitat types, including deserts, grasslands, and even suburban areas where they thrive by dining on lizards when they aren't sleeping underneath landscaping cloth. They can also be found in forests, but are more common in areas with large, arid, open areas suitable for basking. Nightsnakes specialize in two main food items—lizards and their eggs—and occasionally eat other vertebrates. They have rear fangs that inject venom into their prey. Fear not! Nightsnakes are typically too small to bite people. Nightsnakes mate in the spring or early summer, lay eggs in the summer, and hatchlings are born by late summer.

Range and Variations: While I agree with experts that there are two species of nightsnakes in California, I have combined their account in this book because the ecology and appearance of each species are rather similar. The best way to figure out which species you have found is based on location. The Desert Nightsnake (*Hypsiglena chlorophaea*) ranges in a big stripe from southern Canada to northern Mexico, encompassing eastern California, all of Nevada, and parts of surrounding western states. The Coast Nightsnake (*H. ochrorhyncha*) can be found in coastal areas as far north as the Bay Area, plus in the Sierra Nevada.

Marisa Ishimotsu

Coast Nightsnake.

How to Find Nightsnakes: While you can find surface-active nightsnakes by night driving on remote roads through good habitat, most folks have better luck flipping cover objects. Flip any flat-bottomed rocks, fallen logs, or other cover, in just about any habitat, and you have a decent chance of finding these common little snakes. Nightsnakes in particular have a penchant for hiding under extremely tiny objects. I once found a nightsnake sheltering under a cow pie, and now I have to resist the urge to flip every desiccated piece of hooved animal poo that I come across. Nightsnakes sometimes turn up inside buildings, including houses.

You Might Like to Know: Much like juvenile North American Racers, nightsnakes look roughly like baby rattlesnakes. The combination of their blotched pattern, their collar that sometimes makes their heads appear to be a bit wider than their necks, and their vertical pupils, are enough to fool the untrained eye into thinking "rattlesnake!" When threatened, they even flatten their heads out and coil into a rattlesnake-shaped posture. In the case of nightsnakes, it is the adult snakes that are mistaken for baby rattlesnakes. Because nightsnakes seem to thrive in areas with people, like suburban yards, and because they regularly make their way inside homes, I commonly get phone calls from terrified people who have a "baby rattlesnake" inside their house. It is almost always a nightsnake or gopher snake, and occasionally a juvenile racer. I find this "mimicry" of venomous snakes to be fascinating. A nightsnake that had blotches that just happened to be a little more rattlesnake-like was more likely to be avoided by a predator, so it passed on its genes to more babies. Over millions of years of evolution, and millions of such nightsnakes, modern-day nightsnakes may now fool hawks and humans alike into thinking they are rattlesnakes. This is all hypothetical, of course. . . . It has never been experimentally shown that nightsnakes are mimics of baby rattlesnakes. But I have a pretty strong hunch that they are.

Chad Lane

Desert Nightsnake.

Spencer Riffle

CALIFORNIA KINGSNAKE

LAMPROPELTIS CALIFORNIAE

FAMILY COLUBRIDAE

The California Kingsnake is many people's favorite snake. Even people who dislike snakes will admit that this animal is beautiful, noble, and useful. Useful? Below you will find out why most people say that. To me, all of these adjectives apply. Beautiful in the way that their black and white (or sometimes brown and yellow) bands create a mesmerizing apparition when they crawl away. Noble in the way that so many of them seem to patiently accept being held and photographed without biting or musking (Please note that this is not universal. . . . You haven't lived until you have inhaled kingsnake musk on a night drive!). Useful in introducing people to snakes. A kingsnake found by my professor on a field trip was my personal diving board into the pool of herpetology.

Appearance: California Kingsnakes are medium-sized snakes with shiny scales, and in most parts of their range they are banded. This means that the different colors go around their bodies like rings or bracelets, as opposed to stripes that run the length of the snake. The California Kingsnake's bands range from stark black and white to brown and yellowish, and usually extend onto their bellies. That said, kingsnakes with funky patterns have been found in many places in California, and in a few places in the state you can regularly find kingsnakes with different markings. In some areas in the Central Valley, the light bands form narrow and irregular saddles. In coastal Southern California, kingsnakes can be dark with white stripes or spots instead of bands. In most areas, the bands are solidly pigmented, without stipples of other colors inside the bands, although there are rare exceptions to this rule, too. Usually, a kingsnake's nose and lips feature white scales with black borders between them.

Natural History: California Kingsnakes are one of the state's most cosmopolitan snakes, meaning that you can find them in virtually any habitat. You might see one on the beach; you can find them in grasslands and forests of all types up to about 7,000 feet elevation; they are not uncommon in deserts; they are regular visitors of people's yards, even sometimes venturing quite a way into cities. Kingsnakes tend to be active during the day, though they can be found out at dusk or at night when the weather is hot. One notable fact about kingsnakes is that they eat just about *anything*. Their diet includes rodents and other small mammals, lizards, birds, amphibians, and lizard and snake eggs. In addition, about a third of their diet in California consists of other snakes, including rattlesnakes. Kingsnakes kill their prey by constriction. They mate in the spring, lay eggs in the summer, and hatchlings appear in summer or early fall.

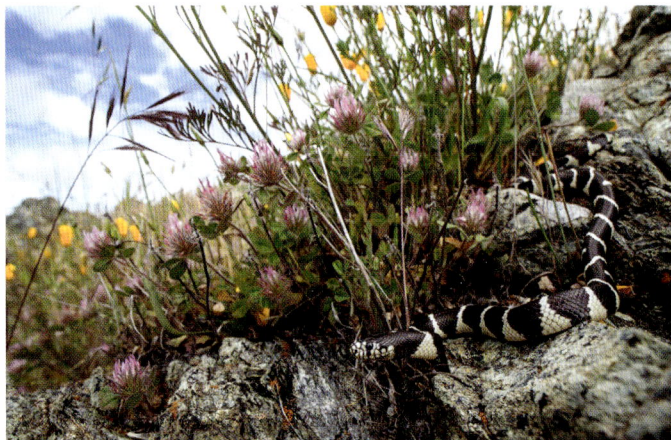

California Kingsnake with typical banded pattern.

Range and Variations: California Kingsnakes can be found most anywhere in California except at very high elevations. Their range also extends into northern Mexico, Arizona, Utah, Nevada, Oregon, and a tiny part of southwestern Colorado.

How to Find California Kingsnakes: Given that the California Kingsnake ranges throughout much of the state, and can be found in practically any habitat, it stands to reason that methods for finding them will also be highly variable. Many kingsnakes are seen cruising through people's yards on the hunt for lizards or other snacks. One of the most common ways that herpers find California Kingsnakes is by turning over cover objects, like pieces of tin and wood, that the snakes like to hide underneath. Night driving can also yield kingsnakes, especially in deserts and other warm areas during the hours shortly after sunset.

Marisa Ishimatsu

California Kingsnake with unusual striped pattern.

You Might Like to Know: The name "kingsnake" arises from the species' propensity for eating other snakes, much as the King Cobra does on the other side of the planet. Above, I referred to the fact that many people find California Kingsnakes to be "useful." This is largely because kingsnakes eat other snakes, including the rattlesnakes that many people consider to be dangerous or pests. While I admire rattlesnakes tremendously and therefore wouldn't personally call kingsnakes "useful" in this regard, one must admit that the ability to catch, kill, and eat a venomous snake that weighs about the same amount as the kingsnake itself is truly amazing! I have observed kingsnakes eating rattlesnakes in both the laboratory and in the field. They try to grab the rattlesnake by the head then rapidly wrap their coils around the snake. Minutes to hours later, the rattlesnake is immobile and the kingsnake ingests it headfirst. Notably, even when California Kingsnakes are envenomated by the rattlesnakes they are trying to eat, they are usually fine because their blood contains proteins that can neutralize rattlesnake venom.

Marisa Ishimatsu

MOUNTAIN KINGSNAKES
LAMPROPELTIS MULTIFASCIATA AND *L. ZONATA*

FAMILY COLUBRIDAE

Widely considered to be one of California's most beautiful snakes, mountain kingsnakes are also the "white whale" for many herpers. Just like Ahab became obsessed with finding Moby Dick, snake enthusiasts spend endless hours driving and hiking around the Sierra Nevada or coast ranges in the hopes of catching a glimpse of that telltale black, red, and white banded pattern. Mountain kingsnakes are not necessarily difficult to find, but they take some know-how, a lot of persistence, and, of course, luck. As their name implies, these snakes are often found at higher elevations, where there tend to be fewer roads bisecting their habitat and where active seasons are often short. The good news is that mountain kingsnakes live in beautiful areas with nice weather and stunning views, making it fun to herp for them whether or not they make an appearance. Herpers who put in the time to find these snakes are rewarded with their stunning beauty as well as with street cred from other herpers. If you do want to flaunt your find, for mountain kingsnakes and any other highly sought after snakes, it is especially important to avoid posting photos that show landmarks or include metadata, lest you unwittingly attract droves of other herpers in search of their own white whales who could trample sensitive habitat en masse.

Appearance: Mountain kingsnakes are medium-sized, narrow snakes with distinctive, shiny red, white, and black bands that encircle their body. The red bands are widest, then white, then the black bands are narrowest. Most mountain kingsnakes have black faces, a white neck, then repeated bands of black-red-black-white down to the tips of their tails. Notably, occasionally individuals can be found that lack red bands and have only black and white bands.

Natural History: Mountain kingsnakes are found in habitat types ranging from scrub to pine forest, and in many places are most common at high elevations. Although mountain kingsnakes are famously found sheltering inside rock crevices, they utilize many habitat types in mountainous regions. Most of their diet consists of lizards and their eggs, though they do eat most any small vertebrate, including rattlesnakes and sometimes even other mountain kingsnakes. Mating takes place after the snakes emerge from hibernation in the spring, females lay eggs in the summer, and hatchlings appear in the late summer.

Range and Variations: There are currently two species of mountain kingsnakes recognized, both of which occur in California. The California Mountain Kingsnake (*Lampropeltis zonata*) ranges from southern Washington, southward through the Klamath Mountains into the northern coast ranges and the Sierra Nevada, and is found from about 1,500 to 8,000 feet elevation. The Coast Mountain

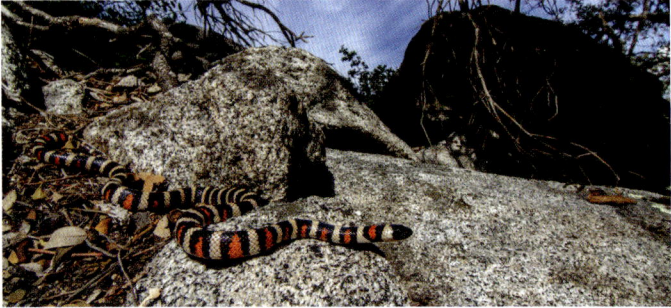

California Mountain Kingsnake.

Marisa Ishimatsu

Kingsnake (*L. multifasciata*) has a spotty distribution in a narrow strip along the coast from Monterey Bay southward to northern Baja California and can be found from sea level up to 9,000 feet.

How to Find Mountain Kingsnakes: Mountain kingsnakes can be found by driving through appropriate habitat, typically during the day when they are most active, although they can be out and about at night when it is otherwise very hot. Although it might seem like it would be easy to spot a red, white, and black snake on the road from your car, this pattern actually can make the snakes surprisingly hard to see, especially if they are crawling across a road covered in shadows of branches and brush. However, experienced herpers know that searching under logs and inside rocky crevices is an excellent way to find mountain kingsnakes. I like to use my cell phone to reflect the sun's rays into rock crevices in search of a glimpse of red. Historically, collectors would use crow bars to separate large pieces of rock so they could capture the kingsnake within; unfortunately, this type of bad behavior persists among some herpers. *Never, ever* use a crowbar or other tool to remove any snakes from hiding spots, as this technique irreparably destroys habitat that is essential to these snakes' survival and may also impact other animals that use those crevices.

You Might Like to Know: The two species of mountain kingsnakes that occur in California are just the tip of the iceberg in this amazing group of animals. Our neighbor to the east, Arizona, boasts two additional species of "tricolor" kingsnakes that are found only at high elevations. But the true tricolor bounty is to be found in Mexico. Multiple species of red, black, and white (and sometimes other colors) beauties occupy remote mountain ranges throughout that country. Scientists debate the exact number of species, partly because these mountain ranges have not been sampled as extensively as those in the United States and partly because different scientists have different opinions on what constitutes a species versus a subspecies or other variety. Mountain kingsnakes are the poster children for the fact that the United States (and sometimes Canada!) represents the northernmost distribution of certain groups of snakes that diversify into Mexico. So, once you have mastered Californian snakes, you can head south of the border and take on the challenge of finding Mexican mountain kingsnakes and so much more.

Marisa Ishimatsu

California Mountain Kingsnake.

Spencer Riffle

COACHWHIPS

MASTICOPHIS FLAGELLUM AND *M. FULIGINOSUS*

FAMILY COLUBRIDAE

The first time I encountered a Coachwhip, I found myself holding out my bloody hands to catch three gooey, partly digested nestling birds as the snake barfed them up. This strange event took place on a quick stop while driving home from a Mojave Desert field trip in my herpetology class when I was a student. Why were my hands bloody? Because I was the one who jumped out of the van and caught the snake as it sped away, and all herpetologists know that Coachwhips almost always bite if grabbed by big, hairy, ape-like predators. Why did it barf? Because the graduate student teaching associate was doing a study on snake diets, so he gently squeezed its stomach to push the contents out, like a tube of toothpaste. There are two species in California—the widespread Coachwhip and the much rarer Baja California Coachwhip that barely crosses the southern border into California. Every herpetologist will say that both species of coachwhip are "cool snakes." Let's delve into why that's true.

Appearance: Both species of coachwhips in California are long and thin-bodied. Their super-long tails have a scale pattern evocative of braided rope. They can be up to 5.5 feet long, which makes them the second longest snake in California after the Gopher Snake. They have huge eyes that are oriented in a forward-facing direction, facilitating excellent vision to pursue prey during the daytime, when they are typically active. The coloration of Coachwhips is highly variable, but is often a mottled brownish, reddish, or pinkish (even sometimes a stunning bright pink or magenta!) with dark and light blotches or bands on the head and neck. Baja California Coachwhips in California are very dark in color, sometimes with a few small white patches along their faces and necks.

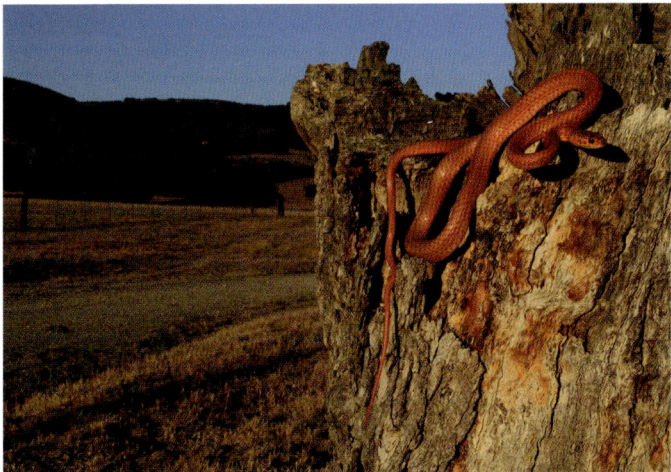

San Joaquin Coachwhip, a threatened subspecies of the Coachwhip.

Natural History: Both species of coachwhip are typically associated with desert habitat and nearby grasslands. They tend to avoid areas with thick vegetation because moving around in this sort of habitat is not conducive to their speedy ways. They are active during the day, even at very high temperatures, although they can occasionally be found active at dusk. At night they typically shelter inside rodent burrows and will escape into these during the day when threatened. They will also readily climb vegetation when fleeing from predators or searching for prey that live in trees. Voracious predators, coachwhips grab their prey and swallow it without using venom or constriction, largely just forcing it down their throats. They seem to be constantly looking for food and they will eat just about any vertebrate (or invertebrates for young snakes) that they can overpower. They even eat roadkill. Coachwhips mate in the spring, lay eggs in the early summer, and hatchlings emerge later in the summer.

Range and Variations: The Baja California Coachwhip (*Masticophis fuliginosus*) ranges throughout Baja California and just barely gets into southern San Diego County where it is designated as a species of special concern. The other species, the Coachwhip (*M. flagellum*), ranges throughout northern Mexico and the southern United States from coast to coast and has two recognized subspecies in California. The more common subspecies, sometimes called the Red Racer, inhabits the southern third of California, along with much of Arizona and parts of Nevada, southern Utah, and northern Mexico. The other subspecies is the San Joaquin Coachwhip and mainly occupies the southern parts of California's San Joaquin Valley, with disjunct populations in Colusa and Sutter Counties. It has a threatened status because the vast majority of the San Joaquin Valley has been developed for agriculture or oil production.

How to Find Coachwhips: Coachwhips are very common in certain parts of California and like to be active outside in the same conditions as us—in warm weather in the morning or midday when we are most comfortable hiking or going for a drive through the desert. Super-active snakes, they may crawl across the trail right in front of you—I once found myself face to face with one while both of us drank from a desert spring in Anza Borrego in San Diego County. I have also regularly found these snakes curled up under cover objects on cool mornings. If you are lucky enough to find a big junkpile in the desert, chances are there is a coachwhip or two hiding within.

You Might Like to Know: Coachwhips have one of the greatest names in all of snakedom. Both its common name ("coachwhip") and scientific name (*"flagellum"* = whip) are nods to the coach-

whip's long, skinny body and especially long tail, which is reminiscent of the braided leather whip that a coachman uses to force his horses to run faster. In general, fast-moving snakes tend to have very long tails. Why is this? I certainly don't know. I posted the question on social media and got numerous suggestions from

Baja California Coachwhip.

Marisa Ishimatsu

the community, including that the long tail might allow more fat storage to fuel movement, might help in making the snake aerodynamic, or provide more muscle to push against the ground. These are all hypotheses that remain to be tested.

Zeev Nitzan Ginsburg

STRIPED RACER

MASTICOPHIS LATERALIS

FAMILY COLUBRIDAE

If you live in or spend a significant amount of time in rural or suburban areas of coastal California or the Sierra Nevada foothills, you will see Striped Racers. They are one of the most common and conspicuous of all Californian snakes. These attractive snakes can be found in many habitats, including people's yards where they feast upon the blue-belly lizards that are so ubiquitous there. Striped Racers are often confused with garter snakes, so make sure that you take a close look before identifying a snake based on the presence of stripes alone. If you decide to try to catch one of these, good luck to you. As their name implies, they are speedsters!

Appearance: Striped Racers are long, thin snakes that are dark brown, gray, or black in color, with a line extending down each side that can be cream, yellow, or orange. They have white spots or stripes on their faces, especially their lips. Their bellies are light in color, sometimes even pinkish. They have huge eyes, indicating that they use vision to stalk prey during the day.

Natural History: These snakes live up to their names by being *fast*. Unless you see them before they see you, you might see only a striped flash disappearing into the brush. While they are commonly seen crawling on the ground or on rocks, Striped Racers are excellent climbers and will readily flee into bushes and trees or climb them in search of prey. Typically found in chaparral, oak woodland, rocky areas, and near water in dry habitats, Striped Racers hunt for lizards by crawling around with their heads lifted off the ground in a distinctive posture. Lizards make up most of their diet, but they also eat other small vertebrates, including snakes and even the occasional rattlesnake. Mating occurs in spring, females lay eggs in summer, and hatchlings appear in early fall.

Range and Variations: The Striped Racer is an example of a Californian snake species with important variations. The common subspecies, the California Striped Racer, occupies the Sierra Nevada up to about 7,500 feet and much of coastal California down into Baja California. The other subspecies, the Alameda Striped Racer or Alameda Whipsnake, has a very limited distribution only in the East Bay. In general, the lateral stripes on the rare Alameda Striped Racer tend to be wider and more orange than those of the common California Striped Racer, and this beautiful orange color extends onto their faces and bellies. Due to extensive urbanization within their already limited range, Alameda Striped Racers are listed as threatened under the California and the United States Endangered Species Acts.

How to Find Striped Racers: Striped Racers are active during the day and tend to like relatively warm weather. I have good luck searching for them by hiking through appropriate habitat at a rapid pace, which means I can cover more ground and increase my chances of meeting one. The key to finding these snakes is to keep your eyes focused as far ahead of you as possible; otherwise, they might see you, courtesy of their excellent vision, and zoom away before you ever spot them. Hiking near streams or small ponds in otherwise dry areas may increase your luck even more. In cool weather, you can find these snakes hiding under cover objects.

You Might Like to Know: Striped Racers are the perfect example of a snake that is way better to watch than to capture. There are several reasons for this. First, much like coachwhips, if you catch them, they are likely to bite you readily and repeatedly. Though these bites are harmless, being bitten by snakes is not most people's cup of tea. More importantly, leaving them alone might earn you quite a performance. Striped Racers are extremely active snakes that always seem to be hunting. If you hang back and watch, you are likely to see them poking their heads around inside crevices and burrows and periscoping their heads up above the vegetation to sneak up on lizards. When they catch the lizard, they will often carry it in their jaws, still with their head up like a periscope, to a safer spot to eat it. They wolf the lizard down, sometimes while it is still kicking, and then go off and look for more. It is much more fun to witness a gluttonous racer stuffing itself on unfortunate lizards than to become a human pincushion.

Francesca Heras

Spencer Riffle

STRIPED WHIPSNAKE

MASTICOPHIS TAENIATUS

FAMILY COLUBRIDAE

I have a confession to make: I have never personally seen a Striped Whipsnake in California. Recently, I learned why I keep missing them. I was in the Mojave Desert with my herpetology students for our annual desert field trip. Another professor on the trip took his students on a drive far up into the mountains to escape the extreme hot and dry conditions, and later told me they had witnessed a Striped Whipsnake eating a mouse. I had been herping that area of the Mojave Desert, rather hard, for twenty-five years, and I had never seen them there! Why? Because they don't occur in the lowland desert. Driving up into the hills that day rewarded those students with a super-lucky encounter.

Appearance: Striped Whipsnakes can grow to be rather long (up to five and a half feet) but are slender to help facilitate their speedy lifestyle. They are usually a dark grey color with thick cream or yellowish stripes along their sides that are often subdivided by little dark stripes. Their bellies are light in color. The scales on the head are dark and often outlined in white, especially at the lips where it gives the appearance of thick white lipstick. Like all racers, Striped Whipsnakes are highly visual daytime predators with large eyes.

Natural History: The Striped Whipsnake is not your typical desert snake. Absent from hot lowland deserts, it is a common inhabitant of sagebrush scrub and woodlands, especially in areas near creeks or ponds. It eats all types of small vertebrates, occasionally raids bird nests to eat eggs or hatchlings, and baby whipsnakes sometimes eat insects. Striped Whipsnakes mate in the spring, females lay eggs in the summer, and hatchlings appear in the late summer.

Range and Variations: In California, Striped Whipsnakes are found in desert scrub communities of eastern California.

How to Find Striped Whipsnakes: Striped Whipsnakes are active during the day, so the best way to find them is by hiking around in scrub habitat. Driving around during the day can work too, but it's often difficult to see them before they see you and take off at top speed. You can also search under cover objects at night where they take shelter.

You Might Like to Know: In the video my colleague took of the Striped Whipsnake with its prey, the snake had grabbed the mouse by its head and was slowly and rhythmically biting down on the head. It was not enough pressure to kill the mouse by crushing its head, just a slow bite down and up. That mouse should have been struggling, but it seemed dazed and confused. So, what was happening? I think that the snake may have been envenomating its prey. Many rear-fanged snakes exhibit this "chewing" behavior to allow the venom from the small fangs to seep into the prey. But Striped Whipsnakes and other racers are not known to have rear fangs or to be venomous. They also don't constrict prey. Most sources just say that these snakes grab prey, "overpower them," and eat them. But based on my colleague's observation, I hypothesize that they have venom that assists them with this overpowering. This would be amazing, if true, because Striped Whipsnakes and other racers are some of the most common snakes in areas where they occur, they have been well-studied, and yet it could be that we have not even discovered that they have venom.

Spencer Riffle

Jonathan Adamski

WATERSNAKES

NERODIA FASCIATA **AND** *N. SIPEDON*

FAMILY COLUBRIDAE

Watersnakes are some of the most fascinating—and stinky—
snakes on the planet! They are famous for the tremendously
odoriferous musk that they eject toward perceived predators,
including humans that pick them up. Once, a colleague took me
to his lab to show me "Big Bertha," a massive female watersnake
he had recently captured and put in a terrarium for a study. When
we arrived, Bertha's terrarium lid was ajar—she had escaped! We
searched all over the lab and he ultimately found her hiding under
the refrigerator. When he yanked her out, she musked him point
blank in the face, literally causing him to run to the sink to vomit.

Jonathan Adamski

Common Watersnake.

Once I stopped pointing and laughing, I caught a good whiff of the stench myself: It smells a lot like dead fish that has been left to rot in the sun for several days, and may just be disgusting enough to make predators drop the snakes and go look for more palatable meals. Well played, watersnakes, well played.

Appearance: These snakes are large (often about 3 feet long) and heavy bodied, usually with a dark background color and lighter brownish or cream irregular markings that are a cross between bands and blotches. A great feature to identify watersnakes are the vertical bars on their upper and lower lips. However, some garter snake species also have these lip bars; when in doubt, take photos and consult a field guide. The easiest way to distinguish the two established watersnake species in California is that the pattern on Common Watersnakes are usually wavy bands, whereas Banded Watersnakes have squarish blotches.

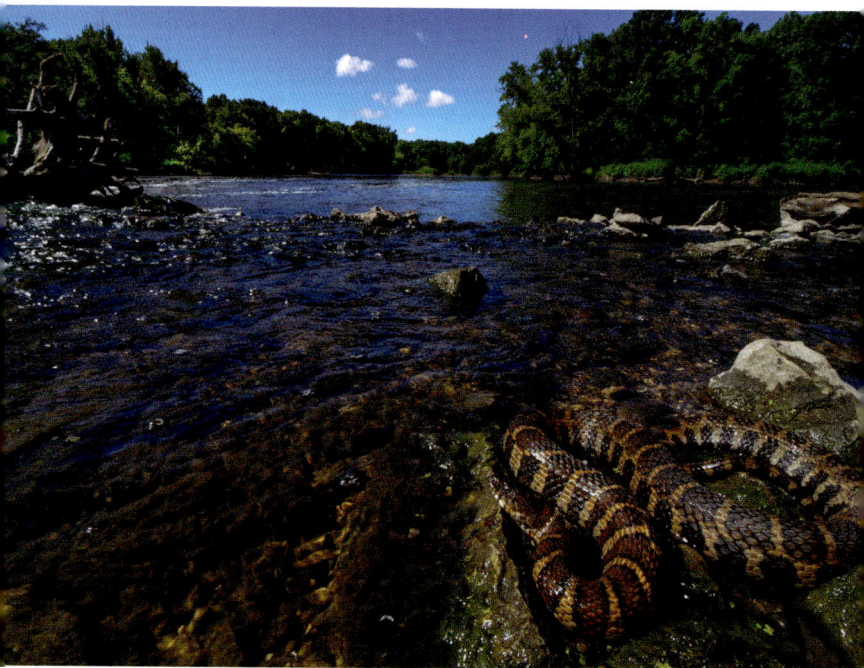

Common Watersnake in its native range in Minnesota. *Photograph by Jeff Lemm.*

Natural History: Watersnakes can be found swimming in bodies of fresh water and basking along their shores, sometimes in large numbers. They are voracious predators that eat most anything they can find, and because they are aquatic, their diet typically consists primarily of fish and frogs. Mating occurs in spring, and females give birth to live young in the summer or early fall.

Range and Variations: Native to the eastern part of the United States, watersnakes are common inhabitants of bodies of water there. In California, two species of watersnakes have become

established over recent decades. Common Watersnakes (*Nerodia sipedon*) inhabit a freshwater marsh near Roseville (Placer County), and a single specimen was found about ten years ago in an irrigation canal near Manteca (San Joaquin County). Banded Watersnakes (*N. fasciata*) have established populations in Machado Lake (Los Angeles County), in several bodies of water near Folsom (Sacramento County), and in the lower Colorado River in extreme eastern Imperial County.

How to Find Watersnakes: The best way to find watersnakes is to walk around the edges of lakes where they are known to occur on sunny days and where you might find them basking on the shore. You can also look for them by kayak or canoe. I don't recommend catching watersnakes—in addition to their impressive stench, they pack a painful (though harmless) bite and are always ready to prove it.

You Might Like to Know: No one knows for sure how watersnakes were introduced to California, but it likely happened when people released unwanted pets into local water bodies. Watersnakes are voracious predators, eating up fish and frogs and possibly outcompeting native aquatic snakes like garter snakes. If you see a watersnake, photograph the snake and post it to the "California Nerodia Watch" page on iNaturalist so that managers at the California Department of Fish and Wildlife can follow up.

Spencer Riffle

SPOTTED LEAF-NOSED SNAKE

PHYLLORHYNCHUS DECURTATUS

FAMILY COLUBRIDAE

For the first part of my career as a herpetologist, I thought that Spotted Leaf-nosed Snakes were rare. I didn't find them during the handful of herping trips I took to the Mojave Desert when I was a college student, and in Arizona we occasionally found them but not regularly. Even when I began teaching herpetology at Cal Poly and started regularly visiting the prime habitat of the Mojave National Preserve, I didn't see any Spotted Leaf-nosed Snakes for the first couple of years. Imagine my surprise, then, when we started seeing numerous Spotted Leaf-nosed Snakes crawling about on desert roads on class field trips! For a string of a few years, they were everywhere. I realized that these snakes

Sean Barefield

are actually one of the more common snakes in that area of the Mojave Desert. Upon doing some reading, I found that Spotted Leaf-nosed Snakes were indeed thought to be rare until Laurence Klauber, widely considered a father of California herpetology, started driving around desert roads at night in the 1920s, where he found them in abundance. It is worth noting that we have once again found very few Spotted Leaf-nosed Snakes in recent, drought-stricken years. As a desert species, it already lives in an area where water is scarce, so further reduction in rainfall can be deadly. Their small size means that loss of water across their skin is accelerated, and the lizards that they rely on for food might also be skipping reproduction due to drought. How Spotted Leaf-nosed Snakes and other desert reptiles respond to long-term fluctuations in rainfall remains to be seen.

Appearance: If you encounter a suspected Spotted Leaf-nosed Snake, the first step to confirming its identity is to ask yourself, "Is it adorable?" If the answer is unequivocally yes, then you may indeed have a leaf-nosed snake. These small snakes are often pinkish or tan with brown spots on their backs and along their sides. They have a very prominent, enlarged, triangular scale on the tips of their noses that gives them a snub-nosed appearance. With some imagination, you can see that the scale is leaf-shaped.

Natural History: Spotted Leaf-nosed Snakes are found exclusively in the desert. They hide under the sand during the day and emerge at night to crawl around the desert in search of their favorite foods, which include lizards and especially their eggs. They use the big scale on their noses to help push their way through the sand and to dig up lizard eggs. Rather little is known about their reproduction, but like many other desert snakes it appears that they mate in the spring, females lay eggs in the summer, and hatchlings appear in late summer.

Range and Variations: In California, Spotted Leaf-nosed Snakes are restricted to the eastern deserts from Inyo County southward, where they extend all the way down to the tip of Baja California. They range eastward into Arizona and northern Mexico and northward into southern Nevada and extreme southwestern Utah.

How to Find Spotted Leaf-nosed Snakes: The single best way to find Spotted Leaf-nosed Snakes is night driving on desert roads in the late spring or on humid summer nights. In certain areas, they can be extremely common.

You Might Like to Know: The Spotted Leaf-nosed Snake is considered to be a dietary specialist. While it eats several types of lizards, it most commonly targets Western Banded Geckos, and especially their eggs. To Western Banded Geckos, the only nocturnal lizard in much of the Mojave Desert, the Spotted Leaf-nosed Snake represents a major threat. These geckos react to the presence of a Spotted Leaf-nosed Snake by arching their backs and waving their fat tails around madly, which directs the snake's attention to the gecko's tail instead of its head. When the gullible snake attacks the tail, the gecko pops its tail right off and escapes. Gecko 1, Snake 0! One study showed that when Western Banded Geckos are exposed to the scent of Spotted Leaf-nosed Snakes, they respond with the tail waving behavior and by running away. Interestingly, when confronted with the scent of Western Shovel-nosed Snakes, which don't typically eat geckos, the geckos skip the tail waving. So, how does a poor Spotted Leaf-nosed Snake ever get a decent meal? It appears that they rely more on the gecko eggs they dig up than on adult geckos themselves. Geckos lay their eggs underground, leave them, and don't stick around to guard them, leaving them vulnerable to being dug up by Spotted Leaf-nosed Snakes. Gecko 0, Snake 1! The balance, of course, is a tie, where Spotted Leaf-nosed Snakes and Western Banded Geckos are locked in a delicate dance of predator-prey dynamics that has, over evolutionary time, shaped their chemical senses and behavior.

Spencer Riffle

GOPHER SNAKE

PITUOPHIS CATENIFER

FAMILY COLUBRIDAE

Sure, finding rare snakes is a major thrill, but there is something extremely impressive about how the incredibly common Gopher Snake has conquered every square inch of California except the snow-covered mountaintops. At certain times of year, I can't leave my house without seeing a Gopher Snake. One of the main things I love about California's most common harmless snake is how wonderful they are for outreach events. At my university, we have three Gopher Snake "ambassadors" that work for their monthly mice by going on elementary school visits with my students. Over the past decade or so, Milo, Snubs, and Bertha have been held, prodded, squeezed, and lovingly petted by many thousands of children. Even the most reluctant child will usually reach out to touch Milo's shiny scales, exclaiming that snakes "aren't slimy" after all! I like to think that our ambassador Gopher Snakes have significantly improved snake public relations among the children

Marisa Ishimatsu

of San Luis Obispo and Santa Barbara Counties over the years.
Positive experiences with snakes like Milo can last a lifetime
and can even trickle up to impact how their families think about
snakes, too.

Appearance: Gopher Snakes grow to be very large, often reaching
lengths exceeding 6 feet. Most individuals you observe will be 5
feet or smaller, though. They are medium in their girth—not as
hefty as a rattlesnake but thicker than racers and other slen-
der-bodied snakes. Gopher Snakes have background coloration
that is typically tan or yellowish but, in some areas, can be tinged
with orange or red. They have a series of squarish dark blotches
along their back that are connected in a pattern that is reminiscent
of a ladder, plus greyish blotches running along each of their sides.
Most Gopher Snakes have a dark stripe that runs diagonally across
the sides of their face, right through their eyes. They often have
a shiny appearance to their skin. Gopher Snakes are commonly
mistaken for rattlesnakes (see page 19), but careful examination
of photos plus a bit of experience with wild Gopher Snakes and

rattlesnakes will help you become adept at differentiating them, even just based on their skin texture and pattern.

Natural History: Gopher Snakes are one of the most common snakes in California. They occur in every habitat type imaginable, from coastal scrub to grasslands and oak woodland, in pine forests in the Sierra Nevada, and into the eastern California deserts. Gopher Snakes can even regularly be found far into urban areas, where they do well moving around among yards and city parks. Gopher Snakes eat a variety of small vertebrates, especially rodents, birds, and bird eggs. Large adult snakes consume mainly rats, squirrels, and rabbits. Mating occurs in spring, females lay their eggs in early summer, and hatchlings appear in late summer.

Range and Variations: Gopher Snakes range throughout most of western North America from southern Canada into northern Baja California and central Mexico. In California, Gopher Snakes can be found everywhere except extremely high elevation regions of the Sierra Nevada.

How to Find Gopher Snakes: If you spend a decent amount of time exploring the outdoors, a better question is how *not* to find Gopher Snakes. They will find you! They are so common in many parts of their range that is often impossible not to see them while out driving or hiking. This is especially true in the spring when male Gopher Snakes are out and about searching for females day and night. You can also find Gopher Snakes night driving, especially in hot areas like the inland deserts. Finally, Gopher Snakes are common users of woodpiles and cover objects of all kinds, so keep flipping boards and searching junkpiles.

You Might Like to Know: Gopher Snakes are among the best climbers of Californian snakes. They can easily climb trees, scale fences, and even climb up the outsides of houses that have textured surfaces like stucco or brick. This skill is central to their notoriety as raiders of bird nests. On a camping trip with my students in the Carrizo Plain, we watched a large Gopher Snake crawl across the campground and climb about thirty feet up a huge tree. As it reached the top, two large birds emerged and started dive-bombing the snake. It was a pair of Great Horned Owls defending their nest! The Gopher Snake completely ignored them and disappeared into the nest. We couldn't see what the snake was doing, but the owl parents' distress as they tried to scare the snake away was a good hint. About forty minutes later, the snake emerged from the nest and crawled back down the tree the same way it had gone up, this time with two large lumps in its belly. In an amazing demonstration of predatory prowess, that snake crawled straight to that nest—it somehow knew it was there. How? I don't know.

Owls regularly feed on Gopher Snakes, grasping and tearing them with their huge talons. Why didn't the owls try to catch and kill the Gopher Snake this time, a move that could have resulted in them feeding the snake to their chicks instead of the reverse? I don't know. One thing I do know for sure: When my students first saw that Gopher Snake crawling across the campground, their natural urge was to run over and pick up the snake to admire it. Instead, we decided to watch the snake to see what it would do. No matter whether you are on Team Owl or Team Snake, the power of observational experiences to connect us deeply to wild creatures in their natural element cannot be overstated. Leave snakes alone and watch them—you never know what you will see.

Spencer Riffle

LONG-NOSED SNAKE

RHINOCHEILUS LECONTEI

FAMILY COLUBRIDAE

When summer days become extremely hot, many creatures start coming out at night. If you go out night hiking, you can see rattlesnakes coiled up like cinnamon rolls, the gleaming eyeshine of wolf spiders, and wary toads springing out of your path. One night I got to witness a remarkable predation event. A Long-nosed Snake was lying in a sandy area with its head poking straight down into the sand. It withdrew its head and then stabbed its face into the sand again nearby. This occurred over and over, as if the snake were bobbing for apples. Finally, it stuck its head into the sand, writhed around a bit, and emerged with a thrashing whiptail lizard in its jaws! It constricted the lizard and chowed it down right in front of

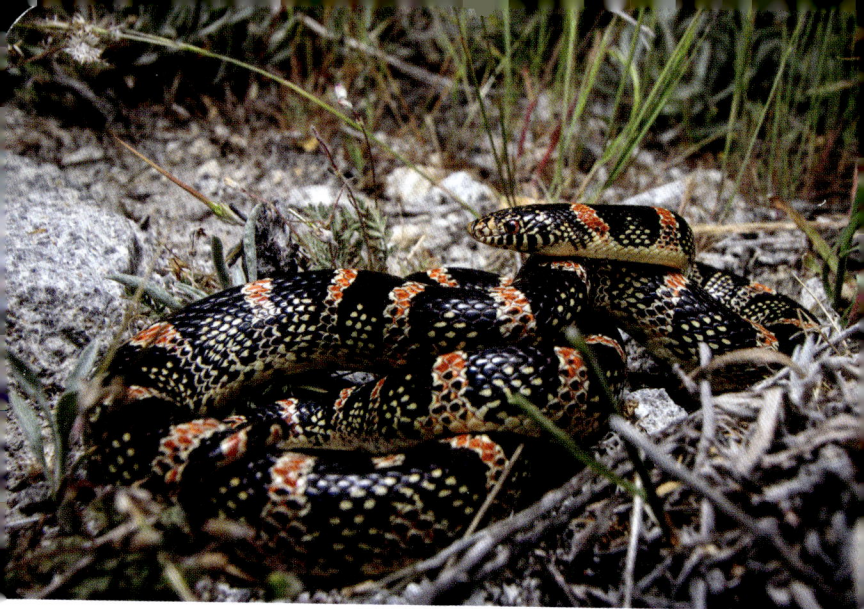
Long-nosed Snake with typical color pattern. Photograph by Zeev Nitzan Ginsburg.

me. As I went about my night, I reflected on how that snake had known that the lizard was sleeping in that particular little patch of sand. Could it sense it moving? Could it smell it? One thing for certain, its namesake long nose helped it pierce the sand easily. If you don't have a shovel, or the hands to hold one, your nose is your best bet!

Appearance: Long-nosed Snakes are medium in size (2–3 feet). They are slender and they have long noses with an underbite. They are always banded, with the bands typically black, red, and white (sometimes just black and white), and the bands don't extend onto the solid cream belly. The bands are usually speckled with the colors from the other bands—for example, black bands speckled with white, white bands speckled with red and black.

Long-nosed Snake with a color pattern that resembles that of a California Kingsnake.

Natural History: Long-nosed Snakes are most common in deserts, grasslands, and dry scrubland, though they can occasionally be found in chaparral or woodland near coastal areas. They are highly nocturnal, though the occasional wayward snake can be found active during the day. Long-nosed Snakes are major predators of lizards, and they also eat lizard eggs and other kinds of small vertebrates. They use their long snouts to dig up lizards and other prey and to burrow in the sand. Long-nosed Snakes mate in the late spring or summer, females lay eggs in the summer, and the eggs hatch in late summer or fall.

Range and Variations: Long-nosed Snakes have a wide geographic range. In California, they occur throughout the deserts, the Central Valley, and the southern part of the state. They approach the coast in Monterey, San Luis Obispo, Orange, and San Diego Counties but otherwise only live in inland areas. They are present only up to about 6,000 feet, so are absent from high-elevation mountain ranges. Range-wide, they stretch eastward into the Midwest and southward well into Mexico, where no one is sure exactly how far south they extend.

How to Find Long-nosed Snakes: Long-nosed Snakes are as night-drivable as they come. On remote desert roads, they can often be the most commonly seen snakes. They seem to be active over a wider temperature range than most desert snakes because you can find them crawling about on hot nights but also on cool nights when other snakes have retreated into burrows. Long-nosed Snakes can also be found hiding under cover objects during the day, though they usually take cover deeper underground in burrows or sand.

You Might Like to Know: All reptiles and birds have a single opening used for urination, defecation, and for sex and egg-laying or birthing: the cloaca. Snakes also have musk glands tucked away in this area, and indeed many snakes spew forth a stinky musk as a defense mechanism when attacked by predators, including unsuspecting herpers who pick them up. You might have had the memorable experience of picking up a garter snake and getting your hands covered in their foul cloacal musk, or even a water-snake whose musk stench reaches legendary status. Well, Long-nosed Snakes have a righteously stinky musk too, but they have a wonderfully fascinating additional weapon in their cloacal arsenal: blood. That's right, when a Long-nosed Snake gets roughed up by a predator, they spew blood and musk out of their cloaca and roll themselves around in it. This is called cloacal "auto-hemor-rhage" because the snakes can seemingly make themselves bleed from the cloaca on demand. Nobody knows for sure what the function of this auto-hemorrhage is, but presumably it startles or disgusts predators, leaving a chance for the snake to escape. It seems to work against herpers, too. Gary Nafis, the creator of CaliforniaHerps.com, summed it up nicely in his account of finding a Long-nosed Snake: "Using a disgusting but effective defensive behavior—it coils up with jerky movements then smears itself with red fluid from its cloaca. After that I certainly did not want to touch the snake again."

WESTERN PATCH-NOSED SNAKE

SALVADORA HEXALEPIS

For many of our California snakes, Southern California is actually the northern edge of their wider range extending down into Mexico. In fact, I found my first Western Patch-nosed Snake not in the state of California, but rather in Baja. When I was an undergraduate, my boyfriend and I went on a big herping trip in the cape region. We couldn't afford to stay at any of the fancy hotels in Cabo San Lucas, so we rented a little posada in the outskirts of town. This ended up being a good thing because we found tons of herps driving back and forth to town on our daily taco run, including the most gorgeous Western Patch-nosed Snake that streaked across the road into a super spiky shrub. We jumped out of the rental car and surrounded the plant. Where is the snake?! I poked and prodded that shrub up one side and down the other, its thorns drawing blood in multiple places, before I finally provoked the snake out of its hiding spot. Up it zoomed, then down, in and out of that thick shrub. This went on for about a half hour before we gave up, tacos calling our names and all that. Though we never caught that snake, I gained a huge appreciation for its incredible speed, climbing skills, and camouflage. Since that day I have met numerous Western Patch-nosed Snakes, most of them racing across the road at warp speed, never to be captured by us pathetically slow humans.

Appearance: Western Patch-nosed Snakes are medium in size (usually about 2–3 feet long), slim, and speedy. They are gray with a wide yellow or tan stripe down the center of their backs, with dark stripes extending from their huge eyes down the sides of their bodies to their tails. The tips of their noses consist of a large, wide scale that they use to dig down into the sand to bury themselves or to uncover tasty lizards that might be hiding within.

Natural History: Western Patch-nosed Snakes are active almost exclusively during the day. They live in dry areas like deserts and chaparral. While they eat many kinds of small vertebrates, they specialize on lizards and their eggs. Western Patch-nosed Snakes mate in the spring, females lay eggs in the summer, and hatchlings appear in the late summer.

Range and Variations: Western Patch-nosed Snakes have a wide geographic range, bound by Southern California, northern Nevada, western Texas, and northern Mexico. In California, they are common in the southern deserts, relatively uncommon in most southern coastal areas, and are occasionally found in northeastern deserts in Lassen County. The Coast Patch-nosed Snake is a

Chad Lane

subspecies that occupies coastal or slightly inland areas from San Luis Obispo County southward into Mexico and is designated as a species of special concern due to habitat loss in its range.

How to Find Western Patch-nosed Snakes: The key to finding Western Patch-nosed Snakes is to cover a lot of ground during the day in the hopes that this maximizes your chances of crossing paths with one on the move. You can get lucky by driving through desert roads from late morning through the heat of the day while keeping your eyes out for speedy snakes crossing the road. Also, flipping cover objects like boards or tin in the early morning or at night can sometimes reveal sleeping patch-nosed snakes.

You Might Like to Know: By now you have probably noticed that the common names of many Californian snakes describe their noses. Patch-nosed, Long-nosed, Leaf-nosed, Shovel-nosed, and so on. This is because many snakes use their noses to do things for which other animals use their limbs. A prime example is digging. The snakes listed above use their noses to dig around in sand or loose soil. Sometimes it's to escape predators, sometimes it's to root around for prey. In the case of the Western Patch-nosed Snake, one of the main purposes of the large scale on their nose is to use it as a shovel to dig up lizards that are sleeping under the sand or to excavate nests of lizard eggs. Ten years ago, my colleague Lisa Hazard published a fascinating natural history note describing her observation of a Western Patch-nosed Snake that used its head as a battering ram to break into a burrow network, then chased a rodent out another hole in that burrow system. This all makes perfect sense—if you don't have hands and feet, you are stuck with only your tail and nose. While tails may occasionally be specialized (I'm looking at you, rattlesnakes), large and reinforced nasal scales are common among Californian snakes.

Ryan Sikola

WESTERN GROUND SNAKE

SONORA SEMIANNULATA

FAMILY COLUBRIDAE

Doing research for this book revealed to me how little is known about certain snakes, and the Western Ground Snake is a perfect example of this. On the one hand, there are hundreds of specimens in museum collections, and this little snake has been the subject of numerous papers on color pattern evolution. However, virtually nothing is known about its natural history. Furthermore, there are big, hulking gaps between the areas where these snakes have been collected or sighted. The snakes are probably there, but when you're talking about a tiny, secretive snake that lives mostly underground, they are bound to go undetected in many areas. I have spent A LOT of time herping in California, and I have only seen two of these little snakes here (although I did see quite a few more when I lived in Arizona). Furthermore, they are so small that studying their behavior using the typical tools available to herpetologists (e.g., radio-telemetry) is not possible. I was shocked when I entered their name into Google Scholar and found almost nothing about their reproduction, diet, and other habits. Like many of the subterranean snakes of the world, the secrets of the Western Ground Snake will remain just that—secret—until engineers of the future create the technology to study them.

Appearance: Western Ground Snakes are also known as Variable Ground Snakes, and the reason for this name becomes clear when you peruse photos of this strange little snake. Typically not much bigger than a pencil, Western Ground Snakes come in a multitude of colors and patterns. In California, most snakes are either banded (and therefore easy to confuse with shovel-nosed snakes, see page 44) or have a colorful stripe down the center of their back. Banded versions might be orange and black or black and white, and occasionally orange and white. Striped versions usually are grey or tan with an orange stripe. Typically, their bellies are a solid tan-grey in color. Additional colorations also occur, but mainly outside of California, including a combined striped and banded variety and a solid brown variety.

Natural History: Western Ground Snakes are a bit of an enigma in terms of their habitat. They are found in many places, from desert flats to grasslands and rocky slopes. However, they seem to appear more regularly in places with some moisture, including areas near creeks and ponds and microhabitats with leaves or other cover that help retain moisture. They spend most of their time underground and may prowl about on the surface when it is raining or at night. Like other small, burrowing snakes, Western Ground Snakes eat invertebrates, including insects and spiders.

Mike Pingleton

Western Ground Snake with banded pattern.

They are thought to produce venom that helps immobilize these prey, but this has not been studied. Western Ground Snakes mate in spring, females lay eggs in summer, and these eggs hatch about two months later.

Range and Variations: Broadly speaking, Western Ground Snakes range from eastern California through the southern United States into Missouri and Arkansas, and northern Mexico. In California, they are found in the eastern deserts.

How to Find Western Ground Snakes: This is a tricky species that may be scarce or locally abundant, but its secretive habits make it difficult to find even in the best of circumstances. You can find them night driving through appropriate habitat on humid summer nights, but you might have better luck flipping cover objects, especially flat rocks. Their records in California are rather spottily distributed on sites like iNaturalist, confirming my suspicion that these geographically widespread snakes are not commonly observed in most areas.

You Might Like to Know: The color patterns of the Western Ground Snake have been studied extensively, with some very interesting results. First, some scientists have suggested that the brightly colored banded pattern might help the snakes avoid predation by mimicking the color of a venomous coral snake, a type of snake that does not occur in California but indeed co-occurs with Western Ground Snakes in Arizona and Mexico. On the other end of the spectrum, those snakes that are solid brown benefit from blending in with the ground. It is rare to see species in which some individuals have strong camouflage and others have strong aposematic coloration (e.g., bright colors that indicate danger, even if it is a harmless mimic of a poisonous or venomous species). These opposing color patterns can even occur within the same populations.

Western Ground Snake with striped pattern.

So, what is the actual purpose of the various morphs? This question has largely stumped researchers. Perhaps the banded pattern of a moving snake helps to confuse predators, as it is known to in other species. The black and red band colors are the result of two different genes instead of being linked together, suggesting that natural selection could act on each color independently. Confusing? I agree. The puzzle of Western Ground Snake coloration is not yet solved.

Spencer Riffle

BLACK-HEADED SNAKES

TANTILLA HOBARTSMITHI AND T. PLANICEPS

FAMILY COLUBRIDAE

You may never see a black-headed snake. Unlike the ubiquitous Gopher Snake, or rattlesnakes that are so dense in certain parts of the state, or even the sharp-tailed snakes that are rarely above ground but are often uncovered when planting tomatoes in the spring, the two species of black-headed snakes are sneaky. Some herpers night drive for years in perfect habitat and never see them, and then on one random evening they might encounter five of them on the road. As a result of their secretive nature, us scientists know very little about them.

Appearance: Black-headed snakes are very small snakes with shiny, brownish bodies and flat, black heads with thin, light collars. Their bellies are often reddish. The two California species are very similar in appearance, and the best way to tell them apart is based on where you find them (see Range and Variations below).

Natural History: Black-headed snakes are found in all kinds of habitats, ranging from deserts to woodland, but they spend most of their time underground in the soil or sand where they hunt small invertebrates like centipedes and scorpions. They inject venom into their prey through tiny fangs located in the backs of their mouths. Their little mouths are too small to bite people, and even if they did, the amount of venom would be miniscule. Because they spend so much time underground, scientists have not observed mating behaviors in the wild, so very little is known. Females lay eggs in spring and summer.

Range and Variations: California has two species of black-headed snakes that look extremely similar. The Smith's Black-headed Snake (*Tantilla hobartsmithi*) occurs further inland, with populations ranging from Fresno and Inyo Counties south to Riverside County, and then eastward as far as south Texas. The more coastal Western Black-headed snake (*T. planiceps*) has disjunct populations that extend from the Bay Area south into Baja California.

How to Find Black-headed Snakes: In the introduction to this book, I discussed how herping is a combination of skill, persistence, and luck. Well, finding black-headed snakes relies on a heck of a lot of the last one on the list. They are undoubtedly abundant deep down in the soil in some areas, hoovering up centipedes left and right. Alas, we humans don't live in the soil, and so we only get a glimpse of these creatures when they venture onto the surface. For this reason, I have only seen about ten black-headed snakes in my life, and each one was a surprise. I have had good luck finding

Western Black-headed Snake.

them night driving on rainy nights, and I also have found several under pieces of tin in the desert.

You Might Like to Know: Because black-headed snakes are so small and nondescript, and because all the species look so similar to one another (not just the two in California, but the numerous species that extend down into South America), scientists rely on certain unusual characteristics to describe and differentiate the species. For black-headed snakes, one of the main characteristics is the shape of the penis. Or penises, I should say, given that all snakes and lizards have two penises, called hemipenes. It turns out that penis shape was a major diagnostic character for many species of reptiles, especially back in the 1900s before genomic sequencing took center stage. Just think of all the hours that twentieth-century scientists sat at microscopes, inflating the tiny penises of snakes with saline and painstakingly recording all their bits and bobs.

Spencer Riffle

AQUATIC GARTER SNAKE

THAMNOPHIS ATRATUS

FAMILY COLUBRIDAE

The next group of species accounts in this book are for the famous garter snakes (genus *Thamnophis*), which you may have heard called "garden snakes" or "gardener snakes." The proper term is actually "garter," which presumably refers to the fact that these thin, dark snakes resemble the garters used to hold up the socks of the men who named snakes back in the day. Regardless of their resemblance to antique clothing accessories, California garter snakes can be difficult to tell apart. They can vary tremendously in color and pattern, and in some cases, you need to count the number of scales on their lips and closely examine the the shape of certain head scales. In this book, we are not going to go into that level of detail. Instead, I will provide a basic description of the appearance

Marisa Ishimatsu

and habits of each species of garter snake, which is often enough to identify the species. Those interested in definitively identifying garter snakes to species can look up detailed information in a field guide or see the online key to identifying Californian garter snakes at www.CaliforniaHerps.com. So, let's begin with the Aquatic Garter Snake, which is one of California's most common—and yes, most aquatic—garter snake species.

Appearance: Aquatic Garter Snakes are medium-sized snakes with a dark background color, a prominent yellow-orange stripe running down the center of their backs, and a light-colored stripe running along each side of the body. They have a tiny, yellow-orange "parietal spot" which is a little dab of color smack in the middle of the otherwise black top of the head.

Natural History: As their name suggests, Aquatic Garter Snakes are highly aquatic, very frequently found in or near water. If you sneak up on one, it will usually flee into the depths of the water. For these reasons, Aquatic Garter Snakes are found in riparian areas like creeks and especially ponds. They eat the typical denizens of such aquatic areas, including small fish, frogs, salamanders, worms, and leeches. In warm areas these snakes can be active year-round, though like most snakes they are more commonly seen from spring through fall. Aquatic Garter Snakes mate in the spring and females give live birth in the summer.

Range and Variations: The Aquatic Garter Snake ranges from Santa Barbara County northward to southern Oregon, mainly in coastal counties.

How to Find Aquatic Garter Snakes: The easiest way to observe Aquatic Garter Snakes is to walk around the edge of ponds or streams during the day, keeping your eyes out for striped snakes. If you see them before they see you, you might be able to sit and watch them hunt, which can be fun. You can also search under logs and other cover objects in areas near ponds, where you could find a snake resting underneath.

You Might Like to Know: Aquatic Garter Snakes are the stars of one of my all-time favorite studies on snakes, along with co-stars Common Garter Snakes, and this study can be likened to reality TV. First, snakes aren't usually considered territorial, in that they don't defend areas or resources like food or mates. They have home ranges, which includes all the areas they use, but they don't usually *defend* areas from other snakes. Not *usually*. Enter the garter snakes. Both Aquatic Garter Snakes and Common Garter Snakes like to eat frogs, which obviously tend to congregate in ponds. Scientists studying wetlands in Monterey County found that when just one of these garter snake species was present, the snakes spent most of their time hunting in the pond and ate lots of frogs. However, if both species were present, the Common Garter Snakes bullied the Aquatic Garter Snakes into fleeing the ponds such that they had to live and hunt in suboptimal spots further away from the water. By bullying, I mean that the Common Garter Snakes would flatten their heads and bite the Aquatic Garter Snakes until they fled. So, it turns out that snakes can occasionally be territorial after all. Leave it to our Californian garter snakes, the drama queens of the snake world, to show us this fascinating scientific fact.

Zeev Nitzan Ginsburg

SIERRA GARTER SNAKE

THAMNOPHIS COUCHII

FAMILY COLUBRIDAE

Californians know that the Sierra Nevada is a special place for biodiversity. Spanning multiple biomes and reaching an elevation of nearly 15,000 feet, this mountain range is home to a wide range of endemic species, which means that they are native to and restricted only to the Sierra. Classic examples include the majestic Giant Sequoia and the adorable Long-eared Chipmunk. There is also an endemic amphibian, the Yosemite Toad. The Sierra Garter Snake is the reptile world's answer to a Sierra Nevada endemic, though it is rarely advertised on tourism websites as an attraction. Like all snakes, garter snakes are typically either vilified or overlooked by the public. While they don't deserve villain status,

I am fine with them being overlooked by most people if it means that they can continue living out their lives terrorizing the fish and the young amphibians of the Sierra. Next time you're up that way in the summer, perhaps to visit one of California's high-elevation parks or to vacation at Lake Tahoe, June Lake, Shaver Lake, or another of our mountain getaways, make it a point to be on the lookout for our endemic mountain garter snake.

Appearance: Sierra Garter Snakes are medium-sized, dark-colored snakes that are rather variable in pattern and overall appearance. They often have checkered markings but these can be faded, sometimes but not always have light-colored stripes down their sides, and occasionally have a light stripe down their backs. They have large eyes and distinct dark vertical bars on their lips. In my opinion, some particularly dark-colored Sierra Garter Snakes rather resemble non-native watersnakes (see page 83), and care must be taken to avoid accidentally misidentifying them.

Natural History: Sierra Garter Snakes are highly aquatic, restricted mainly to ponds and streams from low to high elevations in the Sierra Nevada. They are especially common in and around bodies of water that are studded with rocks and emergent vegetation. They stalk small fish and amphibian larvae underwater and disable them with venom injected from enlarged teeth at the backs of their mouths. Like other garter snakes, this venom is not dangerous to people, although you theoretically could experience mild swelling if you allowed a garter snake to chew on your finger for an extended period, something that many Sierra Garter Snakes would be happy to do. Reproduction in Sierra Garter Snakes has not been studied in detail, but mating likely occurs in spring and then females give live birth to baby garter snakes in the summer.

Marisa Ishimatsu

Fascinatingly, Sierra Garter Snakes have been reported to occasionally reproduce by parthenogenesis, meaning that a female produces babies without any sperm from a male.

Range and Variations: Sierra Garter Snakes are the most "Californian" of all the garter snakes, occurring only in the Sierra Nevada of California plus a small part of its eastern slope in Nevada near Carson City.

How to Find Aquatic Garter Snakes: As you keep reading the garter snake species accounts in this book, you will recognize a pattern. Walk around near ponds and streams in their range, and you are likely to see garter snakes swimming in the water or basking on floating vegetation or at the edges of the water. For Sierra Garter Snakes in particular, I suggest targeting bodies of water with rocks and vegetation in the Sierra Nevada up to about 8,000 feet in elevation. Sierra Garter Snakes are active during the day, so your best bet is to visit the Sierra on a warm summer day. You might wear sports sandals so you can walk around the muddy edges or the shallows of these water bodies, but remember to clean off your sandals with a mild bleach solution between ponds to avoid spreading pathogens.

You Might Like to Know: All garter snakes are good swimmers and enter water to escape from predators, to hunt for prey, or both. However, the extent to which garter snakes are aquatic versus terrestrial varies tremendously among California garter snake species. While Northwestern Garter Snakes and Western Terrestrial Garter Snakes fall a bit more on the terrestrial end of the spectrum, the other species are more aquatic, including the Sierra Garter Snake. It turns out that this tendency is evident in the structure and function of their eyes! Catching prey underwater requires heavy reliance on vision, but water can warp the way that light enters the eye. Sierra Garter Snakes have a very cool adaptation to see better underwater. When underwater, they reduce the diameter of their pupils, which increases the resolution of their eyes so they can better see fish and tadpoles. It's like they have underwater goggles!

Spencer Riffle

WESTERN TERRESTRIAL GARTER SNAKE

THAMNOPHIS ELEGANS

FAMILY COLUBRIDAE

Garter snakes can be difficult to identify to species, even for seasoned biologists. Like me. Yes me, the author of this book on California snakes. I admit that I have trouble distinguishing certain species of garter snakes from one another, especially while watching them from afar without grabbing them. This includes the Western Terrestrial Garter Snake, which can be easy to confuse with Common Garter Snakes in some areas. In defense of us scientists-who-can't-distinguish-garter snakes, there are reasons that it is difficult. First, there is quite a bit of overlap among species in key characteristics. For example, yellow stripes down the back and sides of the body are common to many garter snakes.

Marisa Ishimatsu

Even technical characteristics that we don't go into in this book, like number of scales on the lip, have overlap.

Second, the patterns are highly variable even within one species. Western Terrestrial Garter Snakes are a prime example of this (see below). I am afraid it takes time, experience, and watching a lot of snakes to learn to distinguish them in the place you live, and that the "gestalt" (overall set of features) that you develop for each species might not transfer to other areas. For example, in the Santa Cruz Mountains you can find both Western Terrestrial Garter Snakes and Common Garter Snakes with beautiful red markings down their sides, whereas Western Terrestrial Garter Snakes in inland Northern California and even right down the road in Santa Cruz proper typically lack the red markings. If you see a garter snake with a yellow stripe on its back in the Sierra Nevada, it's probably a Western Terrestrial Garter Snake but could possibly be a Sierra Garter Snake, especially if you are at lower elevations. Don't get frustrated. Think of it as a fun challenge. There may be only eight garter snake species in California, but there is an endless array of beautiful color patterns and markings for you to discover.

Appearance: Western Terrestrial Garter Snakes are medium in size and vary dramatically in color and markings throughout California. They are typically dark in color with light (yellow, orange, or tan) stripes down their backs and along each side. Some populations have red marks and others have dark and yellow blotches that form a checkered pattern along the sides.

Natural History: As its name implies, the Western Terrestrial Garter Snake sits on the terrestrial end of the aquatic-terrestrial continuum in garter snakes. That said, it is still typically found near water, and if startled will escape either into the water or into shrubs or rocks. Across its wide range, it occurs in a multitude of habitats, mainly grasslands and forests, and can be present from sea level to extremely high elevations in the mountains. Western Terrestrial Garter Snakes eat basically anything that moves, including invertebrates and small vertebrates from tadpoles to small mammals. Western Terrestrial Garter Snakes mate in the spring, females are pregnant through summer, and they give live birth to baby garter snakes in late summer.

Marisa Ishimatsu

Range and Variations: The Western Terrestrial Garter Snake is a wide-ranging species, present throughout much of the western third of the country plus southern Canada and northern Baja California. They are absent from Southern California except for a population in the San Bernardino Mountains. They occupy coastal counties from Santa Barbara County northward, plus much of

Northern California including the northern Sierra Nevada and its eastern slopes into Nevada.

How to Find Western Terrestrial Garter Snakes: You can find Western Terrestrial Garter Snakes basking, crawling, or hiding under cover objects in appropriate habitat, especially in areas near creeks or ponds. In the mountains, the best places to look are meadows near significant moisture, like ponds, lakes, or seeps. If you are hiking on a pleasantly warm day in good habitat, you might see a ton of these snakes. They tend to be rather speedy, slinking away into tall grass or occasionally into ponds at your approach.

You Might Like to Know: When snakes have wide geographic ranges, their diets often vary considerably. That makes sense because the available prey are bound to vary, too. But Western Terrestrial Garter Snakes show interesting variation even over short distances, from coastal to inland California. Many coastal populations specialize on eating slugs, while inland populations eat more vertebrates like fish. This preference appears to be inborn. Baby Western Terrestrial Garter Snakes from coastal populations readily flick their tongues and then grab and eat slugs, whereas babies from inland populations turn up their noses at slugs like toddlers presented with broccoli. Adult slug specialists can more easily digest and assimilate slugs than snakes from inland populations can, suggesting some type of physiological specialization in digestion or metabolism. Their teeth are even different: Slug eaters have little ridges along the backs of some of their teeth that might function to help grasp slugs.

Michael Starkey

GIANT GARTER SNAKE

THAMNOPHIS GIGAS

FAMILY COLUBRIDAE

I have never seen Giant Garter Snakes in the wild, I probably never will, and I am fine with that. Why? Because Giant Garter Snakes are a threatened species that is restricted to a small area of California, they live in areas that are difficult to access, and I am happy to let them live their lives without nosy herpetologists chasing them around. To be clear, there is a robust research program aimed at helping conserve these snakes and their habitat, and I mightily applaud those scientists for the work that they do. But the Giant Garter Snake doesn't need *me* (or you) sticking my nose in its business.

Just a century ago, this snake's home, California's Central Valley, was a vast expanse of wetlands and grassland that provided habitat to huge numbers of wildlife, including several species found there and only there, as well as herds of elk, massive flocks of ducks, and, of course, snakes. But a boom in agriculture and oil exploration in the 1940s began a rapid and steady transformation of this habitat, such that it is mostly unrecognizable now. Almost all the natural wetlands in the Giant Garter Snake's historic range have been destroyed, so the snakes must now be largely content to live in artificial wetlands like rice fields.

Historically, Giant Garter Snake habitat would have flooded in the summer months as snow melt from the Sierra Nevada made its way down to the valley, providing important habitat for the snakes and their prey. Some intact natural wetlands have been protected as wildlife refuges in the Central Valley, and in some of these, especially those with year-round water, Giant Garter Snake do quite well. However, even some of those intact wetlands are now drained in the summer and fill with water only in the winter when the snakes are hibernating, so they are not as useful for providing habitat as they could be. Ironically, rice fields in the Sacramento Valley have therefore become so integral to the survival of Giant Garter Snakes that they are prioritized as part of the species' management plan.

Appearance: As you might have guessed from their name, Giant Garter Snakes are big! They are the largest species of garter snakes, typically growing as large as 4 feet but occasionally growing over 5 feet long. These snakes have a dark background coloration and usually have a yellow stripe down the middle of their back and one on each side, with dark, checkered blotch patterns along their sides as well.

Natural History: Giant Garter Snakes are almost always found in the water or sitting on vegetation next to the water. Typical habitat currently includes irrigation canals and rice fields. As aquatic specialists, most of their prey is also found in the water, including fish,

Marisa Ishimatsu

frogs, and tadpoles. Because most native fish and frogs have been extirpated from the wetlands in which Giant Garter Snakes live, much of their diet consists of non-native species like mosquitofish and bullfrogs. Mating occurs in spring, and females give birth to live babies in the summer.

Range and Variations: The Giant Garter Snake occurs only in California. It used to occupy a large area in the lowlands of the Central Valley, but like many other species with that unfortunate historic range, habitat conversion on a massive scale has shrunk its range dramatically. It is currently found in wetlands in the Sacramento Valley from Glenn County southward to San Joaquin County and in the San Joaquin Valley at the intersection of Merced, Madera, and Fresno Counties.

How to Find Giant Garter Snakes: Giant Garter Snakes are a rare and threatened species, so it is not easy to find them. Furthermore, the logistics of their current habitat can make them difficult to observe. You might be able to glimpse one basking on the vegetation on the shores of wetlands in wildlife sanctuaries in the Central Valley. Scientists who study them often use kayaks to move around wetlands, including rice fields and irrigation canals, but these areas are often private. Better yet, just use binoculars to enjoy watching these snakes from a distance so that you do not disturb them. Whatever you do, *do not* capture, touch, or otherwise interfere with a Giant Garter Snake, as this is expressly forbidden due to their threatened status.

You Might Like to Know: Giant Garter Snakes are sneaky. By this, I mean that they often slink into the depths of the water long before you see them. While this is a great behavior for avoiding predators, it also means that scientists looking for Giant Garter Snakes cannot rely on visual searches to detect them. But knowing whether Giant Garter Snakes are present in a given wetland is critically important in managing this threatened species. Many researchers use traps to capture garter snakes, but the sneaky snakes often evade capture. So how is a scientist to know whether snakes live in a certain wetland? Enter a fascinating new technology: environmental DNA (eDNA). The idea is that all aquatic organisms, including garter snakes, shed bits of their DNA into the water (think skin sheds, feces, et cetera).

Scientists recently demonstrated that they can detect the presence of Giant Garter Snakes in water samples from areas with known populations and can even differentiate between Giant Garter Snakes and co-occurring Common Garter Snakes. They then went out and collected water samples from numerous wetlands where Giant Garter Snakes had not been seen or trapped for quite some time and were feared locally extinct, but the eDNA evidence showed that the snakes were still present in over half of these wetlands! I find this study incredibly exciting because it merges biotechnology and conservation biology for the benefit of helping us study species like garter snakes and so many other aquatic species that are at risk due to habitat destruction, pollution, and drought.

Spencer Riffle

TWO-STRIPED GARTER SNAKE

THAMNOPHIS HAMMONDII

FAMILY COLUBRIDAE

About ten years ago, a couple of my graduate students went on a hike at Big Falls in San Luis Obispo County. This is a beautiful, forested spot off the beaten path, with year-round water even in the worst of drought years. They came back exclaiming that the little ponds and creeks were full of California Newts and (invasive) watersnakes! We sat down and looked at their photos, and we realized that the snakes were actually native Two-striped Garter Snakes. Phew! I must admit that the heads of these snakes really look like watersnakes, especially the little vertical stripes on the lips and the large, bulging eyes. However, they are native, and they are fabulous.

Appearance: Two-striped Garter Snakes are medium-sized snakes that come in two main flavors: dark scales with a yellowish stripe or series of spots along their sides, or dark scales all around. That's right- some populations of Two-striped Garter Snakes don't have any stripes. This is a good example of how common names of reptiles can be misleading! The best way to differentiate Two-striped Garter Snakes from other garter snake species in their range is that they lack the light stripes down the center of their backs that most other garter snakes have. The belly is usually yellowish.

Natural History: Two-striped Garter Snakes are strongly associated with aquatic habitats. These snakes therefore eat animals that live in or near these wetlands, including small fish, frogs, and salamanders, plus invertebrates like leeches and worms. They are not constrictors; instead, they inject venom into prey using small

Spencer Riffle

Two-striped Garter Snake without stripes.

Two-striped Garter Snake with faint stripes along its sides. *Photograph by Zeev Nitzan Ginsburg.*

fangs located in the rear part of their upper jaw. Like other garter snakes, this venom is not harmful to people. They mate in spring, females carry the live young, and they give birth in summer.

Range and Variations: Two-striped Garter Snakes are a true Californian species—they range only in California and in Baja California, Mexico. Specifically, they can be found in coastal counties ranging from the Monterey Bay southward.

How to Find Two-striped Garter Snakes: The key for this species is water. The best way to find them is to walk along the edges of ponds or streams and look for them in the water or basking on rocks, but you need to be stealthy because they can zoom way quickly. When I need to collect them for research, I like to wade around in the ponds in shorts or a swimsuit so that I can launch myself at a snake with a mighty bellyflop. However, like other garter snakes, Two-striped Garter Snakes are especially fun to watch rather than catch because they are always doing cool things like chasing fish around. Plus, each time you catch a garter snake, you will be musked with its stinky sauce to within an inch of your life.

You Might Like to Know: The Two-striped Garter Snake is designated as a species of special concern by the California Department of Fish and Wildlife because they live only in certain wetlands in Southern California. Think about it—how many healthy wetlands *are* there in Southern California? We've drained most of them, and drought, pollution, and invasive species are threatening the wildlife in those few that remain, including the Two-striped Garter Snake. To help this snake and its habitat thrive, you can reduce your use of water and encourage others to do so too, and importantly, vote for policy-makers who prioritize protection of water resources for wildlife in California.

Jeff Lemm

CHECKERED GARTER SNAKE

THAMNOPHIS MARCIANUS

FAMILY COLUBRIDAE

Just as many Californian snake species range widely into Mexico, the Checkered Garter Snake is one that is far more common in our eastern neighbor, Arizona, than they are in California. Once, when I was teaching a herpetology class in the legendary Chiricahua Mountains in Arizona, this snake became the star of the show. My students and I arrived at a small wetland in the grassland flats near the mountains after sunset to look for amphibians. We were greeted with a loud chorus of American Bullfrogs, which are non-native to this area and a major threat to native amphibians. A few minutes after we fanned out to look around, a student called out that they had found a snake, and it was eating a frog! We all ran over and found a large Checkered Garter Snake that had hold of a small bullfrog's foot and was trying to eat it leg-first. The notion of a garter snake choosing the frog legs off the menu was hilarious to our group of weary field biologists. The local garter snakes put on an excellent dinner theater production for us, as we found two more gluttonous garter snakes eating invasive bullfrogs that night.

Appearance: Checkered Garter Snakes are medium-sized, have yellow or white stripes down the center of their backs and along each side of the body, and along the rest of the body they have distinctive dark checkered markings on a tan background. Under their eyes they have a dark blotch that is partially interrupted with a tan blotch.

Natural History: Checkered Garter Snakes are typically associated with bodies of water in the desert. They can regularly be found active at any time of day, swimming in or basking on the sides of small ponds, cattle tanks, irrigation canals, and streams. They feed mainly on amphibians but also eat other small vertebrates and invertebrates. Mating occurs in the spring and females give birth to live young in the early summer, but in some populations that have been well-studied in Texas, these females might give birth to a second litter of babies in late summer. About ten years ago, scientists discovered that female Checkered Garter Snakes can either reproduce sexually by mating with a male or can produce offspring all on their own via parthenogenesis.

Range and Variations: Checkered Garter Snakes barely get into the southeastern corner of California, where they inhabit a small chunk of desert in Imperial, Riverside, and San Bernardino

Counties. Much of this range is restricted to the Colorado River that marks the border between California and Arizona, though Checkered Garter Snakes do inhabit a strip of area in central Riverside and Imperial Counties including parts of the Coachella Valley. Outside of California, Checkered Garter Snakes occur from southern Arizona eastward to Kansas and Texas, plus northern Mexico.

How to Find Checkered Garter Snakes: They key to finding Checkered Garter Snakes is searching around desert oases or cattle tanks. During the summer, they are mostly active at night, and I readily find them crossing roads. In addition, if amphibians are breeding when you are night driving in the desert (you'll know it if they are, just turn down your music and listen for their deafening hollers), it is fun to wade through the water and look for Checkered Garter Snakes gorging themselves on the hapless frogs. If you are looking for them when it is less hot out, then you can find them basking on the edges of ponds during the day or swimming around in the water.

You Might Like to Know: Unlike many snake species, Checkered Garter Snakes have been extensively studied by biologists. These studies have ranged from comparison of reproductive traits over a wide geographic range to looking at how a common agricultural herbicide impacts the development of their scales. One particularly interesting topic that has been heavily studied in Checkered Garter Snakes is their chemosensation. All snakes use their tongues to collect chemical information from the air and ground to evaluate their surroundings, a sense that I like to call "smell-taste."

Can garter snakes use smell-taste to detect mates of their species and to find delicious frogs to eat? In one study,

Mike Pingleton

researchers created chemical trails of two species of female garter snakes (one Checkered and one other species) by letting the females crawl across a substrate. They then tested whether male Checkered Garter Snakes could find the trails and distinguish between species. Boy, could they! They always reacted to the female Checkered Garter Snake trail with explosive tongue-flicking, which in snake language means that they find the smell-taste of said trail enticing, and they usually ignored the other species' trail. In another fun study, researchers isolated a protein from frog skin and showed that Checkered Garter Snakes use this protein to smell-taste and then find their favorite meal. All of this makes perfect sense—snakes can't speak, can't really hear, and at night they can't even see well, so their strong sense of smell-taste helps eager Checkered Garter Snakes make their two most important finds, frogs and mates.

Zeev Nitzan Ginsburg

NORTHWESTERN GARTER SNAKE

THAMNOPHIS ORDINOIDES

FAMILY COLUBRIDAE

The Northwestern Garter Snake has one of the smallest ranges in California of all snakes that occur in the Golden State. As its name implies, it is an inhabitant of the Pacific Northwest and just barely dips into the northwestern part of California. Within our state, it only lives on the remote Northern California coast. Herping for this snake is a great excuse for a road trip. And the destination will be spectacular, because Del Norte County has some of the most remote and stunning beaches in the state of California, and because Northwestern Garter Snakes are incredibly beautiful.

Appearance: The folks who named this snake really missed out on an opportunity. They could have called this the Variable Garter Snake, which would have done a much better job of summing up its appearance. This snake usually (but not always) has a wide stripe down its back, and this stripe can be pretty much any color. They also usually (but not always) have stripes on the sides of their bodies. The background color of the skin is also highly variable, and often has rows of dark spots. Their bellies can similarly vary in color and occasionally are covered in little red spots. To complicate matters, there are some populations in which individuals are entirely red or black. Compared to the Common Garter Snake, which is found in the same area of California, the Northwestern Garter Snake is generally smaller with a narrower head and smaller eyes.

Natural History: All garter snakes in California are good swimmers and can be found near water, but the Northwestern Garter Snake is probably the least aquatic of the bunch. If you startle one near the edge of a pond, it is more likely to scoot toward solid ground than toward water. Because they are mostly terrestrial, their diet consists primarily of snails, slugs, worms, along with the occasional fish or amphibian. Mating can take place in both spring and fall, and females give birth to live young in the summer.

Range and Variations: The geographic range of Northwestern Garter Snakes extends from southern, coastal British Columbia, through coastal areas of Washington and Oregon, down into extreme northwestern California.

How to Find Northwestern Garter Snakes: Your best bet for seeing these snakes is to walk around riparian areas during the day and look for these snakes basking near the edges of the water or along rocks further inland from the shore. Northwestern Garter Snakes, like other garter snakes, can also be found hiding under cover objects near ponds and creeks at night.

You Might Like to Know: All garter snakes give live birth rather than laying eggs, and one of the most fascinating discoveries related to this was made in a study on Northwestern Garter Snakes. When snakes have live birth, the embryos grow inside amniotic sacs in the mother's oviducts. Mothers send oxygen to the embryos via tiny placentas and umbilical cords (yes, baby snakes have bellybuttons!). However, the mother doesn't "feed" the developing embryo via the umbilical cord in the same way that humans do. Instead, at the time of fertilization, she deposits a big chunk of yolk into each sac, and the embryos feed on the yolk by pulling it into their bodies via the umbilical cord. A series of elegant studies on Northwestern Garter Snakes showed that mother garter snakes provide their embryos with not just oxygen via the placenta and umbilical cord, but also with water, sodium, and calcium! We now know that this mode of nutrient provisioning is true for all snakes that give live birth, including garter snakes, rattlesnakes, and boas in California, and many other types of snakes worldwide.

Spencer Riffle

Spencer Riffle

COMMON GARTER SNAKE

THAMNOPHIS SIRTALIS

FAMILY COLUBRIDAE

Last but not least in our coverage of garter snakes is the Common Garter Snake. This snake is certainly common. In fact, Common Garter Snakes are one of the most widespread snake species in the world, ranging throughout much of the United States (except for deserts) and Canada, even as far north as the Northwest Territory! This snake also has the distinction of being one of the most thoroughly studied snake species in the world. In Manitoba, Canada, tens of thousands of Common Garter Snakes gather to hibernate underground communally each winter. In May they all emerge to mate, and the forest floor becomes a veritable ocean of snakes—this would be Indiana Jones's worst nightmare, but I bet that most of you reading this book would love to observe this in person.

Marisa Ishimatsu

Appearance: These medium-sized snakes have a single light stripe down their back and one on each side, making them similar in appearance to some other Californian garter snake species. They usually also have a stripe of red spots that extends alongside the light stripes on their sides, which is why they are sometimes referred to by another common name, the Red-sided Garter Snake. All garter snakes have large eyes, but those of Common Garter Snakes are particularly large.

Natural History: Common Garter Snakes are often found near water, but I have found them under cover objects or crawling through the grassland or forest several hundred meters from water. These snakes are active during the day, when they use their excellent vision to hunt down fish, amphibians, and invertebrates. All garter snakes inject venom into their prey with tiny rear fangs, but this venom is not dangerous to humans. Common Garter Snakes mate in the fall and store sperm through the winter, then can mate again in the spring. Females are pregnant during the summer and give birth to live babies later in the summer.

Marisa Ishimatsu

Range and Variations: This extremely wide-ranging species can be found from Canada down to northern Mexico, with multiple subspecies. In California, it can be found in most of the northern half of the state plus along the entire coast. Two subspecies in California warrant discussion due to their conservation statuses. The South Coast Garter Snake ranges along the coast from Ventura down to San Diego and is a species of special concern because most historic populations in this area have been extirpated due to loss of habitat, introduction of exotic predators and competitors, drought, and other human-caused activities. The San Francisco Garter Snake occurs only on the southern San Francisco peninsula. Due to extreme habitat loss in that highly urbanized area, this subspecies is designated as endangered by both the state of California and by the United States.

How to Find Common Garter Snakes: This is the final garter snake species account in the list of Californian species, and if you have already familiarized yourself with *Thamnophis* in this book, by now you have the idea: walk around bodies of fresh water and look for garter snakes basking along the edges, swimming across the surface, or waiting at the bottom to grab unsuspecting fish. If you have fresh water on your property and want to encourage Common Garter Snakes (and indeed, other species as well, depending on where you live), you should lay out small pieces of plywood or tin a few meters away from the edge of the water, in the shade if possible. Chances are you will get a garter snake or two moving in.

You Might Like to Know: The Common Garter Snake exemplifies one of the most fascinating stories of evolution in all of herpetology, indeed possibly even in all of biology. Common Garter Snakes are one of just a handful of predators that have evolved resistance to a toxin present in newts, a type of salamander in the genus *Taricha*. This poison, tetrodotoxin, is a potent neurotoxin that rapidly kills other predators by interfering with signal transmission along their neurons. Common Garter Snakes and several other garter snake species have mutant proteins in their nervous systems that render them largely unaffected when they ingest the toxin. So, these garter snakes have very little competition for this resource! It would be like if suddenly, almost everybody else was deathly allergic to chocolate, so the stores gave all of it away to you. One of the neat things about this evolutionary story is that the various species of garter snakes that are able to eat newts, including Aquatic Garter Snakes, Sierra Garter Snakes, and possibly Western Terrestrial Garter Snakes, each evolved this capability independently, all through random mutations in their nervous system proteins that gave them a "leg up" over the competition.

Zeev Nitzan Ginsburg

LYRESNAKES

TRIMORPHODON LAMBDA AND *T. LYROPHANES*

Every herper thinks lyresnakes are exciting, partly because they are not easy to find in many places, including the eastern Mojave Desert that I frequent. I have found many in the deserts of Arizona, but it took about five years of field trips to see my first lyresnake in California. My students and I were on our annual field trip to the Mojave Desert, where we work ourselves into utter exhaustion with two days full of hiking and night driving and very little sleep. At the end of a long night of night driving that wasn't super-successful, most of the students had given up and nodded off for a nap. Right before we pulled into the field station, I saw it: the unmistakable form of a long, thin snake stretched out on

California Lyresnake.

Zeev Nitzan Ginsburg

the road. I slammed on the brakes and jumped out, the students pitched forward into their seatbelts, and they groggily exited the vehicle to the sound of my jubilant cries as I danced around the prize on the pavement. It was a big, beautiful California Lyresnake. Only later did I discover that no one else had ever reported a lyresnake within forty miles of the area! My herpetology classes now all aspire to find this holy grail of all field trip snakes.

Appearance: Lyresnakes are long (usually 2-3 feet), thin snakes with a markedly wide head that usually bears a large, U-shaped blotch. Lyresnakes have largeish eyes, usually with vertical, slit-like pupils. They are grey or brown with darker blotches, each bearing a pale, horizontal stripe.

Natural History: Lyresnakes are secretive snakes, and there are very few studies on their natural history. They are commonly found in rocky areas in the desert, though they occasionally venture into grassland and woodland. They are nocturnal, and during the day they hide in rock crevices. Lyresnakes eat lizards and occasionally other types of vertebrates, and they inject prey with a toxic venom from fangs located in the back of their mouths, rather different from how a rattlesnake's fangs work. They are not considered dangerous to people, but I would not let them bite and chew on my hand because they could inject enough venom to cause swelling and other unpleasant effects. Reproduction of lyresnakes has never been studied in California, but a study on Arizona snakes suggested that they mate and lay eggs in late spring, and hatchlings appear in late summer.

Range and Variations: A single species of lyresnake was recognized in California until it was split into two Californian species in 2008, and as usual it took quite some time for these new species to be recognized by herpetologists. The California Lyresnake (*Trimorphodon lyrophanes*) is the more common of the two species in the state, occurring in a spotty distribution throughout Southern California into Mexico. The Sonoran Lyresnake (*T. lambda*) is found only in eastern San Bernardino, Riverside, and Imperial Counties and continues eastward throughout the southern half of Arizona into northern Mexico along with small slices of New Mexico and Utah.

How to Find Lyresnakes: In California, you can be in the best habitat, you can have endless energy, and it can still feel like you need to make an offering to the herping gods to find a lyresnake! Lyresnake encounters in some areas of California are indeed largely based on luck with a little bit of know-how thrown in. If you spend enough time out there, it will happen. I suggest hiking around in rocky areas during the day, peering into rocky crevices on the lookout for lyresnakes wedged deep inside them. Then when the sun goes down, get in your car and drive around on roads through rocky habitats. Spring is best in the Mojave Desert, but lyresnakes can be active at night in the summer even when it is really hot and dry. Broadly speaking, lyresnakes in California are more easily found in southern parts of their range like the Coachella Valley than in northern parts like the Mojave Desert.

You Might Like to Know: Lyresnakes have one of the coolest common and scientific names of any snake. A lyre is a string instrument with origins in ancient Greece that is shaped like an upside-down U or the Greek Omega sign. Lyresnakes are readily distinguished from other snakes by the lyre-shaped marking on their head. The genus name *Trimorphodon* is also interesting. It translates into three (tri) shapes (morph) of teeth (odon). Indeed, these snakes have fascinating dentition! In the back of the upper jaw, they have a pair of fangs that delivers venom into their prey. In the middle of the jaw, they have short, straight teeth, the function of which is not known but may create numerous small punctures in the prey to help the venom seep in. In the front of the jaw, they have short, curved teeth that help them pull the (now dead) prey into their mouths, or rather to pull their heads over the prey. The specific epithet *lambda* is a Greek letter that looks like an upside-down V, presumably referring to the blotch on the head. *Lyrophanes* also refers to the lyre-shaped head blotch, of course, plus "phanes" supposedly means "visible." However, I prefer to think of it as being named after the Greek God Phanes, who was a beautiful, winged deity wrapped up in the coils of a serpent.

Scott Eipper

YELLOW-BELLIED SEA SNAKE
HYDROPHIS PLATURUS

FAMILY HYDROPHIIDAE

Sea snakes in California? In recent years, yes. This tropical species appears to be slowly spreading northward, likely because of higher water temperatures associated with global climate change, especially during El Niño years when ocean temperatures are high. Once in 1972 and then several times since 2015, Yellow-bellied Sea Snakes have been found lying on beaches in Southern California. Given that sea snakes are highly venomous, the recent sightings have made the news, where journalists implied that surfers could find themselves competing for waves with hordes of these serpents. This is, of course, totally untrue. The snakes have all been weak or dead, showing that these are merely aberrant snakes being brought northward in warm currents. Sea snake bites to people almost never happen in the water anyway; the few bites that do occur are usually to fishermen handling snakes caught in their nets. The sightings are alarming, however, because they are symbolic of how these high ocean temperatures are having major impacts on marine life in California.

Appearance: Yellow-bellied Sea Snakes have distinct yellow undersides with black or dark brown colors on top. They max out at about 2 feet in length. Their tails are laterally flattened into paddles and are usually marked with black and yellow or white bars or blotches. Their nostrils are high up on their noses to facilitate breathing when the snakes float on the ocean surface.

Natural History: In California, any Yellow-bellied Sea Snakes are in distress, and we should not view anything they are doing (or likely not doing, when washed up on a beach) as being indicative of their normal natural history. In their native range in warm tropical waters, these snakes are typically found way out at sea, where they spend much of their time underwater but can readily be found in floating marine debris that attracts the fish that they hunt. Fascinatingly, all other sea snakes dive to hunt, but Yellow-bellied Sea Snakes pick off fish close to the surface. Like their cobra relatives, Yellow-bellied Sea Snakes kill their prey with highly toxic venom injected via fangs in the front of their mouths. Very little is known about reproduction in this species, except that they mate and give birth to live babies at sea over an extended period.

Range and Variations: Yellow-bellied Sea Snakes are not so much a true California snake as one that has washed up here from time to time. They have the distinction of having the widest natural geographic range of all snakes, inhabiting warm waters throughout the Pacific and Indian (but not Atlantic) Oceans. They are concentrated in tropical areas where ocean temperatures are high.

Scott Eipper

How to Find Yellow-bellied Sea Snakes: Don't bother trying to find these in California. You won't. If you are obsessed with seeing a Yellow-bellied Sea Snake in the wild, fly down to Costa Rica, Panama, or another area in their range if that is closer to you, and hire a boat to take you out to areas with floating marine debris, where it is possible to see dozens of these snakes.

You Might Like to Know: One of my favorite things to do is to read old natural history accounts about snakes. While I realize that this makes me a nerd, I like to think that it makes me a cool nerd, because . . . snakes. I read a wonderful fifty-year-old article describing how Yellow-bellied Sea Snakes literally tie themselves into knots that they then writhe through, scraping one loop of their body through the other. The author hypothesized that this behavior helps reduce fouling (attachment of barnacles and other marine creatures) and assists the snakes in shedding their skin. These were captive snakes, and we do not know if they do this in the wild. I like to think they do, as I find the idea of thousands of sea snakes pret-zeling on the surface of tropical seas very appealing.

Zeev Nitzan Ginsburg

WESTERN THREADSNAKE

RENA HUMILIS

FAMILY LEPTOTYPHLOPIDAE

The Western Threadsnake (a.k.a. Blindsnake) is one of those snakes that only herpers get excited about. Most other people wouldn't even notice it (due to its diminutive size) or wouldn't identify it as a snake (it looks more like a worm). But us herpers, we know a thing or two about why these little guys are quite the find! While not *rare* per se, in most areas Western Threadsnakes are rarely found. They spend a lot of time underground, so finding them on the surface is a real treat. I have found only two of them while herping in the Mojave Desert: one was out crawling on a dry, hot night at the Pisgah lava flow, and the other was underneath a door mat at a research station. Both snakes were greeted with much fanfare, as the opportunity to show off a threadsnake to my students is always welcome.

Appearance: If you are wondering whether you are looking at a worm or a snake, the first thing to do is to look closely to see if

Mike Pingleton

it has scales. Western Threadsnakes are covered in little scales that look like mermaid skin, typically shiny pink, but can also be brownish or lavender in color. This snake is tiny, only up to a foot long and thinner than a pencil. Their "eyes" are dark spots covered in translucent scales, their tail is non-tapered and fat but ends in a point, and they have an overbite.

Natural History: Western Threadsnakes are found in areas in Southern California where they can burrow through sand, ranging from the inland deserts to the coastal beaches. They are nocturnal and may be active on the surface at night, especially on humid evenings or during storms. However, they spend most of their time underground, where they burrow through sandy soils and prey upon termites and ants, including their eggs and larvae. Threadsnakes mate in the spring, and females lay eggs in the summer, which they protect until they hatch.

Range and Variations: Western Threadsnakes range along the California coast from extreme southeastern San Luis Obispo

County and inland from southern Mono County southward into Mexico.

How to Find Western Threadsnakes: Finding Western Threadsnakes in the wild is mostly a matter of luck. Night-drivers with a good eye for tiny snakes on the road will eventually get lucky if they drive enough miles through appropriate habitat, such as low-traffic desert roads though sandy areas. Flipping debris like cover boards, logs, and rocks can also reveal threadsnakes. In my experience with threadsnakes, humid, warm nights are the best time to see them on the road. Because these snakes are so tiny, they can be very difficult to see on dirt roads. Newly paved asphalt roads are ideal because even tiny snakes like these stand out against the substrate.

You Might Like to Know: Threadsnakes are the subject of one of the most wild and wacky natural history stories of all time. Owls sometimes capture threadsnakes and bring them to their nests. Typically, owls kill prey before feeding them to their chicks, but they drop the threadsnakes into the nests alive, where the snakes are not usually eaten by the chicks but rather burrow into the nest material and take up residence within, feeding upon the insects that are attracted to the bits of rotting flesh left over from other meals fed to the messy chicks. The owl-threadsnake relationship is allegedly one of mutualism, where the snakes benefit by getting food and the owls benefit by having personal exterminators that keep insects at bay! One study even found that the owl chicks in nests with threadsnakes grew faster and were more likely to survive than those in nests without these snakes. Further research on this peculiar observation still needs to be done, but it certainly is a delicious story.

Jonathan Adamski

BRAHMINY BLINDSNAKE

INDOTYPHLOPS BRAMINUS

FAMILY TYPHLOPIDAE

The Brahminy Blindsnake is one of a handful of established non-native snake species in California. They got to California (and indeed to much of the rest of the world) via the plant nursery trade by stowing away in the soil of potted plants. When those plants are transplanted into gardens, so are the snakes. So, what is the impact of these non-native snakes? It's hard to say. The Brahminy Blindsnake is so diminutive that its impacts might be on a scale too small to notice, especially when you compare it to notorious invasive snakes like the Burmese Pythons that have taken over the wild lands of southern Florida, eating basically all of the mammals along the way. However, Brahminy Blindsnakes could be outcompeting small, native snakes that also live in urban areas, or it could spread exotic parasites to native wildlife, or it could impact the

Scott Eipper

ecology of urban soils. We won't know until someone does a study on them. What is the scale of their invasion in California? It is also hard to know. Many people, especially southern Californians, likely have blindsnakes in their yards and just don't know it. If you think you have found a blindsnake in your garden, post an image to iNaturalist so scientists can learn more about the growing range of Brahminy Blindsnakes.

Appearance: Like the native Western Threadsnake, the first thing you should do when you encounter a Brahminy Blindsnake (usually by digging it up while gardening) is to look at it closely to make sure it has scales and is therefore not a worm. Once confirmed as a snake, it is rather difficult to distinguish from native threadsnakes. Each species has variable coloration, tails that end in pointy tips, and dark eyespots. However, the eyespot is the easiest way to distinguish the invasive Brahminy Blindsnake, provided that you have really, really good eyesight or a hand lens that allows you see the eyespot in detail. In Brahminy Blindsnakes, the eyespot is toward the top of the scale on which is sits, whereas it is in the center of the scales in native Western Threadsnakes.

Natural History: The Brahminy Blindsnake has spread throughout the globe within soils of plants sold in nursery industries from its native range in Asia. As a result, it is common in people's gardens as well as agricultural areas in Southern California. This is an all-female species (see below) that reproduces without obtaining sperm from males. Although their reproductive biology in California has not been studied, they likely lay eggs in the soil in spring that hatch in summer. Like the native threadsnakes, they eat many invertebrates found in soils including ants and larvae.

Range and Variations: Non-native Brahminy Blindsnakes have been found in gardens from San Diego northward into the Central Valley. They are likely much more widespread than this.

How to Find Brahminy Blindsnakes: As an introduced species, the Brahminy Blindsnake is not typically a species that herpers seek out. Usually, people find this while digging in their gardens.

You Might Like to Know: The Brahminy Blindsnake is the only snake species in the world that consists solely of females. No males exist! Scientists only discovered in the 1970s that Brahminy Blindsnakes are an all-female species. They came to this conclusion when multiple scientists collected large samples of snakes with not a male among them. But perhaps even more interesting was a related discovery that the scientists made when they looked more closely at the snakes they collected. In most species, there is quite a bit of variation in physical traits (like color, number of rows of scales, et cetera). However, within populations of the Brahminy Blindsnake, there was almost no variation. This lack of variation combined with the discovery that they are all females led scientists to hypothesize that Brahminy Blindsnakes reproduce via parthenogenesis, where females produce fertile eggs without sperm from males.

Also interesting is the fact that Brahminy Blindsnakes are triploid, meaning that they have three copies of each chromosome. In contrast, other snakes are diploid (two copies) just like you and me. Triploidy suggested that the Brahminy Blindsnake arose when two diploid, sexually reproducing species hybridized in the past, producing a triploid offspring that happened to be very successful in that environment. Indeed, these snakes have success in spades. Part of their extreme success in spreading around the world arose from the fact that they can reproduce whenever they please, with no need for finding males. If this works so well, then why don't other snakes just reproduce by parthenogenesis instead of by finding a mate? First off, scientists are discovering that many sexually reproducing snake species do occasionally reproduce by parthenogenesis (including some Californian garter snake species). However, the Brahminy Blindsnake is the only one that exclusively uses parthenogenesis. Although blindsnakes appear to be doing great right now, the lack of genetic variation means they could be susceptible to extinction if conditions change unfavorably for them. Indeed, over evolutionary time, sexually reproducing animal species do tend to stick around longer than all-female species.

Zeev Nitzan Ginsburg

WESTERN DIAMOND-BACKED RATTLESNAKE

CROTALUS ATROX

Western Diamond-backed Rattlesnakes are near and dear to my herping heart. Some experienced herpers call Western Diamond-backed Rattlesnakes "s**t snakes" because they are so common in southwestern deserts, including the southeastern corner of California where they are abundant. I couldn't disagree more. I studied them for my dissertation research as a PhD student in Arizona back in the 2000s, where on hot summer nights we would encounter droves of these snakes crossing the road or coiled in the sand. Not only are they beautiful, charismatic, and play very important roles in the ecology of the desert Southwest, but their success as a species is extremely noteworthy. Western

Zeev Nitzan Ginsburg

Diamond-backed Rattlesnakes are tough. They eat a lot, grow quickly, reproduce readily and frequently, and seem impervious to certain things that other snakes cannot stand. For example, construction of homes and farmland drives many snake species away into wild areas, but Western Diamond-backed Rattlesnakes thrive in the agricultural areas of the Imperial Valley. This resilience is, of course, a good thing or a bad thing depending on your point of view. Most farmers and residents don't want rattlesnakes in their fields or yards. Luckily, the large (and growing) network of people who volunteer to remove rattlesnakes from properties can help by relocating rattlesnakes, thereby rescuing both the snake and the property owner.

Appearance: Western Diamond-backed Rattlesnakes grow to be very large (up to about 7 feet, though chances of seeing one larger than 4 feet are extremely low in California). They are also thick and heavy snakes, in fact the heaviest of all Californian snakes. Their background color ranges from pinkish tan to brown, with darker diamond-shaped blotches on their backs that are usually outlined in white. They have a distinct "coontail" with black and white bands that are approximately equal in width, and like all rattlesnakes, their tail ends in a rattle.

Natural History: Western Diamond-backed Rattlesnakes are quintessential desert snakes in California, although they can also be found in agricultural fields that have been planted in our deserts. Their activity follows the weather; in spring and fall they are often active during the day, whereas in summer they are out in early morning and evening, and become nocturnal when it is extremely hot during the day. Young snakes eat small vertebrates including lizards, birds, and small rodents, and as they grow they switch to larger mammals like rats, squirrels, and rabbits. Like all rattlesnakes, they inject venom into prey with two large fangs located in the front of their mouths, then follow the scent trail left by the prey on the ground until they find it and ingest it. Western Diamond-backed Rattlesnakes mate in both the spring and fall, and females give live birth to baby rattlesnakes (called pups!) in the summer or early fall. Like most rattlesnakes, females don't usually give birth every year, but instead it takes one or more years between births for them to gain enough weight to support another pregnancy. However, in this species, if they get enough food they can produce pups multiple years in a row.

Range and Variations: Western Diamond-backed Rattlesnakes inhabit only the extreme southeastern part of California (parts of Imperial, Riverside, and San Bernardino Counties, with one snake recently found just barely inside San Diego County) and extend eastward all the way to Arkansas and southward to southern Mexico.

How to Find Western Diamond-backed Rattlesnakes: In prime desert habitat in their range, Western Diamondbacks are very easy to find. This is because they tend to be numerous and, as ambush hunters, they curl up above ground rather than hiding inside burrows for much of the day. Hike through desert habitat, especially rocky areas in spring and fall and lowland washes in summer. In the heat of the summer, hike at night with an excellent flashlight or headlamp, looking for the telltale "cinnamon roll" appearance of a rattlesnake coiled up waiting for a rodent to run by. You can also find snakes on the move by night driving through desert roads on hot summer nights, especially humid ones. On milder days, look under desert plants or rock piles where rattlesnakes like to shelter.

You Might Like to Know: Given the success and resilience of Western Diamond-backed Rattlesnakes, encounters with people are very common. While numerous advocacy groups and snake relocators are working hard to help promote peaceful coexistence of people and snakes, rattlesnakes are still mercilessly persecuted by many people. This is especially true for Western Diamond-backed Rattlesnakes, which die by the thousands at annual rattlesnake roundups that have been going on for decades. Luckily, California doesn't have any of these barbarous events, but they are held annually in many communities in some southern states.

The most famous of these roundups occurs in Sweetwater, Texas, annually in March. The dubious star of their show is the Western Diamond-backed Rattlesnake, which comprises the majority of the snakes brought in by people competing to capture the most snakes or the biggest snake. In 2021, over 3,600 pounds of rattlesnakes were brought in, where they were then decapitated and skinned alive (their bodies continue to wriggle and their heads gasp for breath for hours after decapitation) in front of huge crowds. No permits or other regulations are required, and participants are allowed to pour gasoline into snake dens, forcing out the inhabitants and poisoning the land in the process. In case you can't tell, rattlesnake roundups nauseate me.

Luckily, Western Diamond-backed Rattlesnakes are a tough species, and they are not in danger of extinction due to such unmitigated destruction, though certainly some populations have declined due to overharvest. Even more luckily, as people's opinions about rattlesnakes change, more and more rattlesnake roundups are being converted into rattlesnake festivals, where crowds are still attracted to benefit the economy of the community, but snakes are not harmed in the process. This past spring, the last remaining rattlesnake roundup in Georgia was converted to a rattlesnake festival. When I congratulated the organizers on their Facebook page, one unhappy resident responded angrily, "Are you one of those outsiders trying to change the way people think about snakes, changing our way of life?" I proudly responded, "Yes. Yes, I am."

Spencer Riffle

SIDEWINDER
CROTALUS CERASTES

The Sidewinder rattlesnake, or just "Sidewinder" if you prefer, can melt the hearts of even the most snake-averse among us. Its diminutive size, the high-pitched buzz made by its tiny rattle, its mesmerizing "sideways" locomotion, and the little horn-like structures over its eyes that give it a constant expression of surprise all combine to make this little rattlesnake one of the most endearing of all Californian snakes. I still remember the first Sidewinder I ever saw. I was taking herpetology at UC Berkeley in the 1990s, and we visited the Kelso Dunes in the Mojave National Preserve for a night hike. As the sun set over craggy granite peaks to the west, groups of students wandered around the massive silvery, windswept sand dunes in search of nocturnal creatures. Someone hollered "Snake!" in the distance. All around me, flashlight beams paused, then rapidly started bobbing in a single direction as everyone raced to see the find. I arrived to find students crouched around a creosote bush at the edge of the dunes, peering intently at the tiny, coiled figure half buried in the sand at the base of the shrub. Pretty much every student who arrived emitted a "Squee!" or "Ermagerd!"

Since that day twenty-five years ago, I have brought hundreds of my own students to the same dunes to observe Sidewinders cratered in the sand, and the students' reactions are always of absolute delight at witnessing this iconic desert snake. Finding them on the road while night driving is similarly exciting because we get to witness their sidewinding locomotion as we usher them off the road to safety. Sidewinding is hard to describe in words and must be seen to grasp, but basically the snake moves loops of its body in a sideways motion such that only certain parts of its body are in contact with the ground at any one time. This behavior is thought to be an adaptation to locomotion on sand because it minimizes contact with the hot substrate and because it efficiently allows the snake to move across soft and shifting surfaces without slipping. It is surprising how fast a Sidewinder can move in this way when it finds itself illuminated in the flashlight beams of eager herpetology students.

Appearance: Sidewinders are the smallest rattlesnakes in California, with adults typically being about 1 foot long and rarely exceeding 2 feet. Their coloration is a light tan or pinkish-white with darker blotches, a pattern that helps them camouflage in the sandy areas that they inhabit. Sidewinders have dark stripes extending from their eyes sideways onto their cheeks, and the scales above their eyes are elevated into characteristic horn-shapes. No one knows the function of the horns for certain, but they might protect the eyes from strong solar radiation or sand. Their rattles are small even in proportion to their body size, so when they rattle in warning the noise resembles a high-pitched buzz. Some people who have age-related hearing loss cannot hear the high-pitched rattle at all.

Natural History: Sidewinders are found in sandy, desert habitats. Like most rattlesnake species, young Sidewinders eat primarily lizards and add rodents to their diet when they get larger. Because they are such a small species, Sidewinders often continue eating lizards throughout their lives. They use venom to kill their prey, typically biting and releasing then following the scent trail of the envenomated prey and ingesting it after it has died. Sidewinders typically mate in the spring and give birth to live young in the late summer. Unlike most other rattlesnakes, female Sidewinders often give birth annually instead of skipping years.

Range and Variations: Sidewinders can be found in sandy habitats throughout the southeastern deserts of California, as well as southern Nevada, southwestern Arizona, northern Mexico, and a tiny corner in southwestern Utah.

How to Find Sidewinders: Sidewinders are one of the easiest snakes to find in California. Look for sandy desert areas on a map (for example, using the satellite image layer on Google Maps) and hike around during the active season (spring through fall) at dawn, dusk, or at night. Sidewinders prefer sandy areas with vegetation; for example, at the Kelso Dunes you are more likely to find them near the creosote bushes that circle the dunes than in the sand-only areas deep within the dune complex. Another great way to find Sidewinders is to night drive on desert roads that transect sandy areas. On warm spring nights in the Mojave Desert, I have found over ten Sidewinders in a single hour by night driving in good habitat. Remember that Sidewinders, like all rattlesnakes,

Marisa Ishimatsu

are dangerously venomous and should never be handled except by experienced, trained individuals.

You Might Like to Know: If you were to travel to the Sahara Desert of northern Africa, to the Namib Desert of southwestern Africa, or to the Arabian Desert in the Middle East, and if you were to hike around in sandy habitats, you might be surprised to encounter small, horn-eyed species of vipers that move across sand via sidewinding, but that are not closely related to Sidewinder rattlesnakes! It turns out that this suite of physical and behavioral characteristics evolved convergently in several desert vipers, which means that they are look and behave similarly not due to shared evolutionary ancestry but rather because they evolved in similar habitats that exerted parallel natural selection pressures. Even microscopic structures on the belly scales of all these species convergently evolved to help these snakes move across sand. I don't know about you, but I find examples like this to be among the most exciting in the entire field of biology because, for me, they really show how and why certain structures and functions evolve. Viva sidewinding!

Marisa Ishimatsu

WESTERN RATTLESNAKE

CROTALUS OREGANUS

FAMILY VIPERIDAE

Choosing a favorite snake for me is like asking a parent to pick a favorite child—nearly impossible! The expected answer might be the most striking, beautiful snake, and on some days sinuous snake beauty is indeed the most compelling draw for me. However, as a biologist I have a more nuanced view of what it is to like—or indeed, to love—wildlife. Beauty matters, sure. But more important is the feeling I have when I encounter a snake in the wild. It is the thrill of the hunt, and the greater thrill of finding a snake. It is watching their interesting behaviors, which I might have missed if I had interfered with them by prodding them or capturing them. It is their success in persisting in high numbers even in areas where people persecute them. And this braid of connection is strongest for me with the Western Rattlesnake, which I confess is my absolute favorite snake. I have uncovered hundreds under tin boards at junk piles. I have radio-tracked hundreds with my students over the years, getting a chance to peer into their private lives for months or even years. I have rescued hundreds from yards where, if not for me, animal control officers or homeowners might have beheaded them with shovels just for daring to exist. With familiarity comes understanding and affection.

The Western Rattlesnake, like other rattlesnakes, is unfairly vilified and sensationalized by nearly everyone. The truth is that this species, which is notable in California because its wide range and high density means that it is the rattlesnake species that most often comes into contact with people, is actually a gentle snake that wants nothing to do with us. In fact, it does us a great service by controlling rodent populations that, if left unchecked, would denude our plants, eat our crops, and spread disease. Western Rattlesnakes can live for decades, have complex social lives complete with friends and family members, and give live birth to pups that they protect from predators during the vulnerable period before their first shed. The sheer volume of my experience with these beautiful snakes has bred in me a strong affection and an urgent need to share what I know about living peacefully with Western Rattlesnakes.

Appearance: Western Rattlesnakes are medium to large snakes, typically 2-3 feet but occasionally as large as 4 feet. They are thick-bodied like other rattlesnakes, and of course possess the conspicuous rattles on the end of their tails. In terms of color, Western Rattlesnakes vary tremendously. Their background color can be tan, brown, greenish, yellow, or orange, with dark

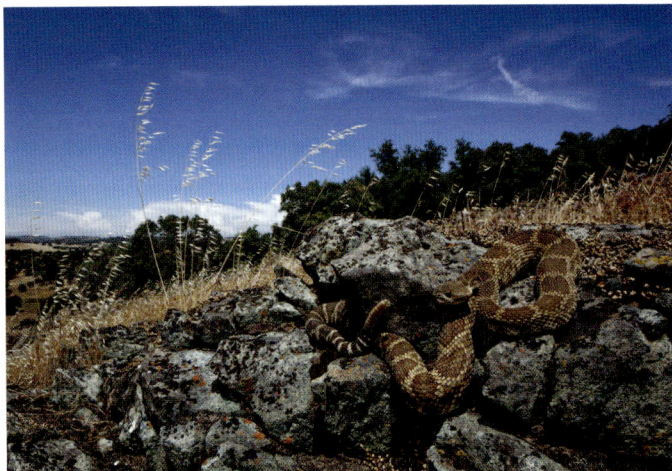

Northern Pacific Rattlesnake, a subspecies of the Western Rattlesnake.

blotches on their backs that are often outlined in a lighter shade. However, some snakes are very dark, even completely black, and occasionally people come across a light-colored rattlesnake with no pattern. Western Rattlesnakes pups are light in color with dark blotches and are born with a "button" on their tail that is often yellow. They will instinctively shake this button, which is rather comical because it does not make a noise. Only after two rounds of shedding will the pups have produced a new rattle segment that can make noise when it vibrates against the button.

Natural History: Western Rattlesnakes inhabit every habitat in California except for Southern California deserts, and even there they range ever so slightly into desert edges here and there. They live in sandy, vegetated beaches on the coast, in oak woodland and chaparral, in and around agricultural fields and orchards, in the grasslands of the Central Valley, and up into the pine forests to well above the tree line of the Sierra Nevada, where in some areas they range as high as 12,000 feet. Western Rattlesnakes use their venom to kill rodents including mice, rats, and squirrels, but also occasionally birds and amphibians, and as babies they eat mainly lizards. California Ground Squirrels constitute an important part of the diet. Like other rattlesnakes, Western Rattlesnakes have heat-sensitive "pits" on their noses that help them detect and strike warm prey, like rodents, even in total darkness. In inland populations, Western Rattlesnakes mate in the spring and fall, whereas mating can happen practically year-round in some coastal areas. Females are pregnant in the summer usually every 2–3 years and give birth to live young between July and October. Sometimes female Western Rattlesnakes gather at a rock outcrop to gestate, give birth, and care for their pups in a communal nest called a rookery.

Range and Variations: There is vigorous debate among scientists and naturalists as to whether the variants of the Western Rattlesnake constitute subspecies or species. There are three variants that occur in California. The Great Basin Rattlesnake occurs from Mono County up through Lassen, Del Norte, and eastern Siskiyou Counties and outside of the state occupies the Great Basin Desert of Oregon, Idaho, Utah, and northern Arizona. The Northern Pacific Rattlesnake ranges from southern Canada down to San Luis Obispo and Santa Barbara Counties. Southward from the central coast is the Southern Pacific Rattlesnake, which ranges in coastal counties south into Baja California.

How to Find Western Rattlesnakes: Western Rattlesnakes are the perfect snake to look for on hikes. On mild days they are often active above ground, typically coiled in the shade of a plant or rock or seen on the move. In the spring, look for rattlesnakes hiding in large, complex rock outcrops. If you are in an area with large populations of California Ground Squirrels, peer into the mouths of their burrows. Western Rattlesnakes love to hide in a good junk pile as much as any other snake, and given their high population densities, you have a high probability of finding one sheltering under a piece of tin or a wood board. Finally, you can find Western Rattlesnakes on the move by driving around during mild weather on roads that bisect good habitat.

You Might Like to Know: Western Rattlesnakes are the stars of the one of the greatest stories in all of biology. The saga concerns the Western Rattlesnake and a species with which its life is intimately intertwined, the California Ground Squirrel. Rattlesnakes eat the squirrels, and also use their burrows for shelter from the elements and predators. Given the diminutive size of a ground squirrel, it might shock you to know that the adults can sometimes survive a bite from an adult rattlesnake. This is because the squirrels, over millions of years of coevolution, have evolved enough resistance to the venom of rattlesnakes (in the form of proteins that bind up the venom and reduce its activity) that they can sometimes survive. This evolutionary dance of ratcheting up attacks and then defenses was also accompanied by behavioral changes.

Adult ground squirrels are bold and will harass, taunt, and even attack and kill rattlesnakes that have come too close to the squirrel's burrow. Ground squirrels that have spotted a potential snake predator do a specific behavior where they rapidly wave their tails back and forth in the air as they stare at the snake. This behavior likely has many functions, but one of them is particularly gasp-worthy. The tail waving behaviors directed to a Western Rattlesnake and to other potential snake predators appear the

Great Basin Rattlesnake, a subspecies of the Western Rattlesnake. *Photograph by Chad Lane.*

same to us. However, scientists filming the behaviors with an infrared heat-sensing camera discovered that the squirrel tails become very hot *only* when they are confronting rattlesnakes, apparently accomplished by shunting warm blood into the vessels of their tail. The result is that the tail glows like a lightsaber in the face of a rattlesnake, which can sense infrared radiation using its heat-sensing pits! This apparently works, as the rattlesnakes tend to slink off and hunt elsewhere when squirrels harass them with the tail waving display. The coolest part of this is that the squirrel *doesn't* bring the heat when facing down other snakes like Gopher Snakes, somehow instinctively knowing that those snakes wouldn't be able to visualize the infrared. What an incredible story of coevolution!

Marisa Ishimatsu

SOUTHWESTERN SPECKLED RATTLESNAKE

CROTALUS PYRRHUS

FAMILY VIPERIDAE

I am going to do something unusual and share a herping spot with you. The secret has been out for a long time, after all, so it's only fair that you know about it, too. The Pisgah lava flow lies alongside Hwy 40 east of Barstow in the Mojave Desert. It is an epic spot for herping, not least because many of the herps there are dark as a result of natural selection for blending in with the lava. While Sidewinders and Mojave Rattlesnakes live in the sandy areas and creosote flats that surround the lava, the Southwestern Speckled Rattlesnakes are the kings of the lava itself. They are rock specialists, and lava is volcanic rock. If you visit Pisgah when it is super-hot out (which it quite often is in the desert), you might see black lizards darting around the lava, but you probably won't see Southwestern Speckled Rattlesnakes. However, if you herp there when temperatures are mild, the place is crawling with some of the most beautiful "specks" (as herpers call them) I have ever seen. Dark and marbled with black, grey, white, and orange, the Pisgah "Specks" are a sight to behold.

Appearance: Southwestern Speckled Rattlesnakes are medium-sized snakes that typically have a pattern that is—you guessed it—speckled in appearance. They do have the blotches on their backs that are typical of Californian rattlesnakes, but these are often shaped like narrow saddles that tend to fade into adjacent background colors in a stippled pattern. Southwestern Speckled Rattlesnakes come in every color you can imagine, with a propensity to blend into the colors of the habitat in which their population has evolved. Their heads are particularly wide and triangular, and their thick tails have dark and light bands, though these aren't usually quite as distinct as the pure black and white "coontails" observed in Red Diamond or Western Diamond-backed Rattlesnakes.

Natural History: Southwestern Speckled Rattlesnakes tend to be found in rocky areas. These venomous snakes eat primarily rodents, but they will occasionally eat birds and lizards. In most areas of their range, Southwestern Speckled Rattlesnakes engage in mating activity in the spring and early summer. Females ovulate in late spring, are pregnant throughout the summer, and give birth to live babies in the summer. In some areas of their range, they give birth every year in a row, but in other places it can take them two or three years to gain enough weight to support pregnancy.

Range and Variations: Southwestern Speckled Rattlesnakes are primarily a desert species. They inhabit the Mojave Desert of California and southern Nevada and extend into the Sonoran Desert of Central Arizona and the Baja California Desert extending northward into the coastal San Diego area.

How to Find Southwestern Speckled Rattlesnakes: Since Southwestern Speckled Rattlesnakes are often associated with rocks, you should night drive through desert roads that transect rocky areas, especially on warm spring nights when males are most active as they search for mates. Hiking in the early morning, at dusk, or at night in the hot summer can yield Southwestern Speckled Rattlesnakes hunting for prey, often coiled among rocky outcrops or in adjacent flatlands. They are typically nocturnal, though you can find them coiled at the base of rocks or plants in the early morning hours or sometimes even during the middle of the day on cool days. Like all rattlesnakes, searching under cover objects like pieces of tin or wood or large plant leaves in appropriate habitats can also reveal hiding Southwestern Speckled Rattlesnakes.

You Might Like to Know: Seasoned biologists and snake enthusiasts know a secret that many beginning herpers don't know yet: Southwestern Speckled Rattlesnakes are clearly the most beautiful snakes in the world. Of course, beauty is in the eye of the beholder, but just behold this snake and dare to disagree. The color palette of Southwestern Speckled Rattlesnakes is unmatched among all Californian, even potentially all American, snakes. In volcanic flows in the desert, like Pisgah, they are much darker than in neighboring sandy areas, sometimes even black. But "Specks" also come in pink, red, orange, brown, grey, bluish-white, and more. We

Marisa Ishimatsu

treasure bright colors and beauty in nature, like monarch butter-flies and California poppies, and surely these visually stunning creatures will strike even those who dislike snakes as beautiful. The Southwestern Speckled Rattlesnake can help show even snake skeptics that snakes are beautiful animals and worthy of our protection. This is especially important for rattlesnakes, thousands of which die at the ends of shovels each year because people think they are ugly, scary, disgusting, or dangerous. Try it! Show this photo to someone who doesn't like snakes and challenge them to tell you he is not beautiful. You may just help change their mind.

Spencer Riffle

RED DIAMOND RATTLESNAKE

CROTALUS RUBER

FAMILY VIPERIDAE

A huge, beautiful, brick red-colored Red Diamond Rattlesnake was the very first rattlesnake I ever saw in the wild. I was driving up Tramway Road outside of Palm Springs with another student in my biology class, on our way to capture Granite Spiny Lizards for a study (or *try* to capture them . . . but that is a story for another book), and we slammed on our brakes to avoid hitting this big, beautiful rattlesnake, with a bright black and white banded coon-tail, that was moseying its way across the road. On that day back in the 1990s, I had no idea that my first wild rattlesnake was such a special snake. Red Diamond Rattlesnakes are the only protected species of rattlesnake in California, designated as a species of special concern by the California Department of Fish and Wildlife. This is largely because the species only ranges into a few counties in Southern California, with most of its range in Baja California. The rapid urban development occurring in much of their range in Southern California likely has a disproportional impact on their populations. Red Diamond Rattlesnakes tend to have rather gentle temperaments, making them a wonderful rattlesnake to search for and observe in the wild.

Appearance: As their name suggests, Red Diamond Rattlesnakes are often red in color, ranging from pink to bright red to dark red, and sometimes brownish. They have somewhat diamond-shaped blotches, and their tails have distinct black and white bands (similar to those of Western Diamondbacks). These snakes get very large, up to 5 feet long. They are heavy bodied, with large individuals easily being as big around as a person's forearm. Like most rattlesnakes, the babies have a bolder pattern, with the blotches standing out more prominently against the lighter background color.

Natural History: Red Diamond Rattlesnakes are commonly found in chaparral and scrub habitat, especially in areas with large rocky outcrops. They can also be found at the edges of woodland and deserts near rocky scrub. These rattlesnakes are often active during the day in the spring and fall when temperatures are mild and at night in the summer when it is hot. On the coast, this species can be active year-round, but in colder inland areas they usually hibernate during the winter. Red Diamond Rattlesnakes primarily eat rodents and rabbits, occasional birds, and smaller snakes eat lizards. They ambush their prey and kill them by rapidly injecting highly potent venom through two hypodermic needle-like fangs. Red Diamond Rattlesnakes mate in spring, females are pregnant throughout the summer, and they give birth in the late summer and early fall. Females usually only become pregnant every 2–3 years, or even less frequently.

Range and Variations: The Red Diamond Rattlesnake ranges from extreme southeastern Los Angeles County, south though Orange and San Diego Counties, and into western San Bernardino, Riverside, and Imperial Counties. They have a tiny geographic overlap with Western Diamond-backed Rattlesnakes in central Riverside and southern San Bernardino Counties.

How to Find Red Diamond Rattlesnakes: Finding Red Diamond Rattlesnakes isn't particularly hard if you are in great habitat. Hike around rocky areas at times of day when temperatures are mild, looking for snakes tucked up under rocks or coiled underneath shrubs. If you can find good junk piles or cover objects in appropriate habitat, you are likely to get lucky. You can also drive around in the morning or evening on roads in good habitat hoping to see one crawling across.

You Might Like to Know: Because Red Diamond Rattlesnakes are designated as a species of special concern, it is illegal for people to harm or kill this species. I don't believe that *any* rattlesnakes should be killed for trespassing in people' yards, but killing a Red Diamond Rattlesnake is especially problematic because their populations are already under extreme stress due to the massive amount of development taking place in their habitats, especially for coastal populations. Because of this, Red Diamond Rattlesnakes have been the focus of several studies on the efficacy of relocation to mitigate human-wildlife conflict.

Zeev Nitzan Ginsburg

In general, most research suggests that short-distance relocation (moving snakes within their home range, usually less than half
a mile) is better for snakes than moving them far away, where they
might wander around and get eaten by a hawk or run over by a car.
This is also true for Red Diamond Rattlesnakes. Notably, however,
two recent independent studies on Red Diamond Rattlesnakes
have shown that snakes relocated longer distances also do well.
While this is good news because it suggests that these sensitive
rattlesnakes could be relocated instead of simply killed when
their habitat is slated for development, the relocation of snakes
is not simple and straightforward. Instead, protecting habitat in
the already heavily developed areas of their range, especially in
Orange and San Diego Counties, and conducting outreach for
people living in areas with these snakes to convince them that
they are in fact gentle giants that merely want to be left alone to
eat rodents, are more desirable goals than trying to solve all rattlesnake problems by relocating them.

Dave Zeldin

MOJAVE RATTLESNAKE

CROTALUS SCUTULATUS

FAMILY VIPERIDAE

Nothing is more iconic of the American Southwest than a Mojave Rattlesnake. As their name suggests, in California this species of rattlesnake is found almost exclusively within the Mojave Desert. It does not occur anywhere near coastal California, and yet when I give presentations about rattlesnakes, audience members from up and down the California coast swear that the big rattlesnake they saw in their yards was a "Mojave Green." I consulted on a National Geographic television program called *Something Bit Me*, and in both rattlesnake bite cases they featured, the first responders mistakenly thought the bites were from "Mojave Greens" even though they were not within the geographic range of this snake. Hysteria, rumor, and exaggeration plague this rattlesnake in California.

No, Mojave Rattlesnakes are not aggressive and they do not chase people (the absence of any footage showing such behavior in today's camera-ridden culture should confirm that). No, Mojave Rattlesnakes are not interbreeding with other species of snakes in California, spreading the genes for their toxic venom into other species (at least not much). No, Mojave Rattlesnakes neither get 8 feet long nor produce litters of 125 babies. These myths and many more are summarily busted in my colleague Mike Cardwell's book, *The Mohave Rattlesnake: And How it Became an Urban Legend*, which I highly recommend. So, what is going on with the people in my central coast neighborhood who swear that the rattlesnakes they see locally are Mojave Greens? They are actually Western Rattlesnakes (*Crotalus oreganus*, see page 166), which simply have variable color patterns, including many with a greenish hue. I have encountered many hundreds of Mojave Rattlesnakes and Western Rattlesnakes in my life, and I can confidently say that both species are beautiful, impressive, gentle when left alone, spirited when harassed, and overall extremely cool snakes.

Appearance: Mojave Rattlesnakes are medium in size, with most individuals 2–3 feet in length. Some individuals have a greenish hue, but many are light brown or yellowish. Mojave rattlesnakes have a black and white banded tail like the "coontails" of Red Diamond and Western Diamond-backed Rattlesnakes, though in Mojave Rattlesnakes the black bands are usually much narrower than the white bands. Many experienced herpers can easily distinguish Mojave Rattlesnakes by their blotches, which have a

Zeev Nitzan Ginsburg

unique jagged shape outlined in darker brown. Finally, between the
scales over their eyes they have only two large scales that touch
one another, whereas other species have many smaller scales (but
I don't recommend getting close enough to see this!).

Natural History: In California, the Mojave Rattlesnake is a desert
species, as its name suggests, although it can also be found in
adjacent grassland and scrub. Mojave Rattlesnake activity tracks
the weather: They are usually nocturnal but during mild weather
can be found active during the day, especially early in the morning or at dusk. Most of the desert that these snakes inhabit gets
cold in the winter, so they typically hibernate underground during
winter months. Mojave Rattlesnakes use their highly toxic venom
to kill just about any type of vertebrate, including toads, lizards,
and birds, and most of the adults' diet is small rodents. They mate
in the spring and in the late summer and fall; females become

Zeev Nitzan Ginsburg

pregnant in spring using sperm from a recent mating or sperm stored from previous matings, and give birth during the summer. When ample food resources are available, female Mojave Rattlesnakes might have babies annually, but in more stressful times like prolonged drought, the snakes can forgo breeding until they have gained enough weight and found enough drinking water and food to successfully reproduce.

Range and Variations: The Mojave Rattlesnake's range falls squarely within the Mojave Desert proper: They occur in eastern Kern County, northern Los Angeles County, extreme southern Inyo County and northern Riverside County, and in most of San Bernardino County. Note that their name is sometimes also spelled Mohave Rattlesnake.

How to Find Mojave Rattlesnakes: I recommend driving desert roads at dusk until a few hours after sunset on warm spring nights. The best roads are those that transect flatlands with extensive creosote bush, as Mojave Rattlesnakes are highly associated with these plants. You can also hike around the desert in the morning before temperatures get too high, looking for rattlesnakes coiled up in the shade of creosote bushes. If you happen upon a junkpile that is not on private land, jump for joy at your luck and then carefully search under the cover objects for Mojave rattlesnakes.

You Might Like to Know: All rattlesnakes are misunderstood, perhaps none more so than the Mojave Rattlesnake. Their defensive behavior is exaggerated and twisted by television programs, YouTubers, reporters, your neighbor who swears he was chased by an "8-foot Mojave Green," and just about everyone else. Mojave Rattlesnakes are commonly said to be "more aggressive" than other species, a statement that is mere lore and unsubstantiated by facts. Their venom is also said to be far more toxic than other California rattlesnake species. While Mojave Rattlesnake venoms indeed often contain neurotoxins in addition to the protein-digesting enzymes present in all rattlesnake venoms, the statement is overly simplistic and frankly irrelevant in most contexts.

Neurotoxins can be found in some other rattlesnake species in California; venom composition varies dramatically among individual snakes, not just species; bite victims' physiological responses to envenomation are at least as important to the bite symptom severity as the type of snake that bit them; and *all* rattlesnake bites are an extreme medical emergency that require immediate care, with all rattlesnake envenomations requiring the exact same treatment protocol. Finally, bites from Mojave Rattlesnakes are relatively rare in California, likely because encounters between people and Mojave Rattlesnakes are uncommon in the sparsely populated desert. So, as disappointing as this is to hear for people who feed off the dramatically overblown image of rattlesnakes as huge fiery-venom-breathing beasts, the truth is that all rattlesnakes, including Mojave Rattlesnakes, never attack people or their pets, and instead only bite to defend themselves.

Chad Lane

PANAMINT RATTLESNAKE

CROTALUS STEPHENSI

FAMILY VIPERIDAE

My encounters with Panamint Rattlesnakes have occurred at the southern part of their range, where I take my herpetology students annually to explore the Mojave National Preserve. My first time teaching the class, I was leading a night hike with my students in a lava flow in the northern reaches of the preserve. We had just seen a California Kingsnake and a Desert Nightsnake on the move during our hike, and on the drive over we had seen several delightful Sidewinders, so spirits were high. The "herping gods were smiling on us" is one common way we spoke about such lucky nights. Well, the herping gods had more in store for us that night. I still recall the rush of adrenaline when I saw the round

Chad Lane

shape tucked up under a pile of rocks in the glow of my lantern. "Rattlesnake!" I hollered. As my students ran down the wash to join me, I reflected on one of the main reasons I love herping for rattlesnakes so much. They follow rules. Southwestern Speckled and Panamint Rattlesnakes love rocky areas in the deserts, and if you spend enough time herping these areas during good conditions, you will encounter one.

Appearance: Panamint Rattlesnakes are medium-sized rattlesnakes, reaching about 3 feet in total length. Their pattern is made up of bands that extend onto the animals' sides and often appear jagged around the edges, although sometimes these look more like roundish blotches. Their tails are usually banded, and typically the furthest dark band appears to join directly with the dark base of the rattle. Their color is extremely variable and is often similar to that of the substrate in the area due to natural selection for

Marisa Ishimatsu

camouflage. Often, but not always, they have a dark greyish or even bluish hue to their cheeks.

Natural History: Research on the closely related Southwestern Speckled Rattlesnake abounds, though few studies have focused on the specific habits of Panamint Rattlesnakes. Like their speckled cousins, these rattlesnakes are typically found among rocks, although occasionally they can move away from rocks into grassy areas to hunt for rodents. Like most other rattlesnakes, their activity tracks the weather, and they can be active day or night depending on the season. They are dietary generalists that eat small vertebrates like lizards and rodents, using venom to kill prey before swallowing them. Unlike some other species of rattlesnakes that mate in both the spring and fall, in Panamint Rattlesnakes the mating activity appears to be restricted to spring. Females are pregnant during summer and give live birth to live babies in late summer. Although annual reproduction might be theoretically possible, Panamint Rattlesnake females typically produce pups every three years on average.

Range and Variations: Panamint Rattlesnakes became their own species in 2007 and were recognized as a subspecies of the Southwestern Speckled Rattlesnake before then. Panamint Rattlesnakes have a small geographic range, extending from the desert of east-central California into southwestern Nevada.

How to Find Panamint Rattlesnakes: Rattlesnakes in California are all rather closely related (they all belong to the same genus), so it makes sense that they share a number of natural history traits and behaviors. So, if you are reading the rattlesnake species accounts in this book in order, you are likely seeing a pattern emerging. Similar techniques (hiking and night driving through appropriate habitat) are successful for finding all California rattlesnake species. Like Southwestern Speckled Rattlesnakes, you can find Panamint Rattlesnakes by night driving through desert roads that transect rocky areas. Hiking or searching under cover objects in rocky canyons can also be highly effective.

You Might Like to Know: The Panamint Rattlesnake is a relatively new species. Not new evolutionarily, per se, but newly split from other speckled rattlesnake species. In Eastern and Southern California, Panamint Rattlesnakes and Southwestern Speckled Rattlesnakes share similar habitats and habits but diverge enough for scientists to consider them separate species. Diverge in what ways, though? And just how much divergence is enough to be considered different species? Well, the answers to these questions vary among scientists and change over time as technology improves. When I was a college student, we still learned about the biological species concept, which was the notion that species are those that can interbreed with one another, whereas if members of separate species try to interbreed, they either will fail or the offspring will be sterile. It is now known that many species hybridize in nature, and rattlesnakes are no exception. Instead, a species is now defined as a group whose members exhibit a certain level of similarity in morphological (physical) characteristics and in their DNA, while subspecies are variations within a species. While Panamint Rattlesnakes were initially a subspecies based on morphology data, the addition of genetic data showed that they diverge enough from other speckled rattlesnakes to be named their own species. Specifically, they diverge in both the sequences of their nuclear DNA and of their mitochondrial DNA, showing strong support for the Panamint Rattlesnake to be named its own species.

ACKNOWLEDGMENTS

I am thankful to so many people who helped make this book a reality. First and foremost, huge thanks the amazing photographers who allowed me to use their photographs, including Marisa Ishimatsu, Jonathan Adamski, Brittany App, Sean Barefield, Scott Eipper, Zeev Nitzan Ginsburg, Francesca Heras, Chad Lane, Jeff Lemm, John Perrine, Mike Pingleton, Spencer Riffle, Ryan Sikola, Michael Starkey, Tanner Statham, Joshua Wallace, and Dave Zeldin. Several colleagues read early versions of the book and gave me valuable feedback, including Brian Hinds, Brandon Kong, and Joshua Parker. Bob Hansen is a treasure trove of technical information about California snakes. I also owe thanks to Gary Nafis, who is the creator of the comprehensive online field guide CaliforniaHerps.com. I have used that site to gather facts about California snakes for over fifteen years, and I highly recommend it to continue learning about snakes in more detail. For inspiration, I credit Carl Kauffeld's 1957 classic, *Snakes and Snake Hunting*, which inspired me to hit the field in search of snakes, my undergraduate mentor Harry Greene, who encouraged me numerous times to write a book, and Marthine Satris of Heyday, who emphatically agreed that the world needed a book on California snakes. Thank you to my family who has always been supportive of every endeavor I have taken on, especially my wonderful husband, Steve, who endures my endless chatter about snakes, all day every day, always with a smile on his handsome face.

RECOMMENDED FURTHER READING

Are you excited to learn more about snakes? Here are a few recommendations for you.

FIELD GUIDES

Hansen, Robert W. and Shedd, Jackson D. *California Amphibians and Reptiles (Princeton Field Guides)*. Princeton University Press, 2025.

This excellent new field guide features high-quality photographs of all California amphibian and reptile species along with detailed information on identifying features and natural history.

Stebbins, Robert C. *A Field Guide to Western Reptiles and Amphibians (Peterson Field Guides)*. Houghton Mifflin Harcourt, 2003.

This older field guide is hard to find and has some taxonomy that is now out of date, but its beautiful and intricate paintings of Californian amphibians and reptiles make it a classic that is worth adding to your library.

CaliforniaHerps.com

I highly recommend this excellent free online field guide to the amphibians and reptiles of California, managed by Gary Nafis.

BOOKS

There are many, many books on snakes. Here are four of my favorite books on the scientific aspects of snakes.

Graham, Sean P. *American Snakes*. Johns Hopkins University Press, 2018.

Greene, Harry W. *Snakes: The Evolution of Mystery in Nature*. University of California Press, 1997.

Lillywhite, Harvey B. *How Snakes Work: Structure, Function and Behavior of the World's Snakes*. Oxford University Press, 2014.

Steen, David A. *Secrets of Snakes: The Science Beyond the Myths*. Texas A&M University Press, 2019.

SCIENTIFIC LITERATURE

I have curated a list of the scientific studies that I consulted when writing this book. It is available on my website at EmilyTaylorScience.com.

ABOUT THE AUTHOR

Brittany App

Emily Taylor is a professor of biological sciences at the California Polytechnic State University in San Luis Obispo, California, where she conducts research on the physiology, ecology, and conservation biology of lizards and snakes with her students. She has a bachelor's degree in English at UC Berkeley and a Ph.D. in biology at Arizona State University. This is her first popular science book, though she has been a biology textbook author for many years. Self-described as "obsessed with snakes," Emily is a staunch advocate for improving the public image of snakes, especially rattlesnakes. She founded a community science project called Project RattleCam, where members of the public help her and other scientists learn about rattlesnakes by analyzing photos and livestream footage from snake dens (rattlecam.org). She also founded a company, Central Coast Snake Services, aimed at helping people and snakes in California coexist safely and peacefully (CentralCoastSnakeServices.com). She lives in Atascadero with her husband, Steve, and their menagerie of rescue animals, including Pax the dog, Baby the Boa constrictor, Aperol Spritz the bearded dragon, and rattlesnakes, Buzz and Snakeholio. Follow her on social media @snakeymama and access her website at www.EmilyTaylorScience.com.

Common Sharp-tailed Snake.
Photograph by Spencer Riffle.